SOME THEORETICAL PROBLEMS OF CATALYSIS

SOME THEORETICAL PROBLEMS
OF CATALYSIS

Research Reports of the 1st Soviet-Japanese Seminar on Catalysis

Edited by

Takao KWAN

Georgii K. BORESKOV

Kenji TAMARU

UNIVERSITY OF TOKYO PRESS

© UNIVERSITY OF TOKYO PRESS, 1973
    UTP 3043-67872-5149

Published by
UNIVERSITY OF TOKYO PRESS

PREFACE

In 1971 the Scientific Council on Catalysis of the USSR Academy of Sciences and the Catalysis Society of Japan decided to hold annually a Soviet-Japanese seminar on catalysis in the USSR or Japan alternatively in order to discuss the current problems concerned with catalysis research. The first seminar was held in July 1971 to debate particularly the theoretical problems of catalysis with 20 participants from both countries at the Institute of Catalysis of the USSR Academy of Sciences in Novosibirsk. This volume consists of 18 articles presented at the seminar and 4 session discussions. In addition to papers included in this volume, many important communications by Soviet scientists were presented and discussed at the seminar. We regret that these communications and discussions could not be included in this book because of page limitations.

The term "some" in the title of the book was decided upon as this volume does not contain all possible theories of catalysis. One purpose of this book is to introduce a phase of catalysis research going on in both countries to researchers, engineers, and students all over the world in the hope that it would contribute to the advancement in this field.

We would like to express our sincere gratitude to those who presented English manuscripts in spite of the difficulties of writing in English, a language which is not a mother tongue for either country. Finally, we wish to thank the University of Tokyo Press for publishing this volume.

August 1973

The Editors

CONTENTS

Preface
Introductory Remarks
..............................G. K. Boreskov   3
Mechanism of the Selective Oxidation of Olefin by Metal Oxides
.... Y. Moro-oka and Y. Takita   7
Optimal Bond Energies with Catalyst of Reactants and Catalytic
   Reaction Products
..............................G. K. Boreskov   23
On the Activities of Oxide Catalysts for the Allylic Oxidation of
   Olefins
.... T. Seiyama, M. Egashira and M. Iwamoto   35
On the Mechanism of Oxygen Activation during Carbon
   Monoxide Catalytic Oxidation on Silica-Gel Supported
   Vanadium Pentoxide
...... M. J. Kon, V. A. Shvets and V. B. Kazansky   51
Discussion   59

Electronic Interaction of Molecular Oxygen with Metal
   Porphyrins in Solution
.... K. Yamamoto and T. Kwan   73
Description of the Interaction Effects of Chemisorbed Particles
   Based on the Surface Electronic Gas Model
.... M. I. Temkin   83
Catalysis by Electron Donor-Acceptor Complexes: Mechanism of
   Hydrogen Exchange Reactions on Electron Donor-Acceptor
   Complexes of Aromatic Compounds with Alkali Metals
.... K. Tamaru   91
Experimental Investigation of the Contribution of Charge Transfer
   from Complexes to Catalysts
.... O. V. Krylov   105
Mechanism of Carbon Monoxide Oxidation on Manganese Dioxide
   Catalysts as Revealed by a Transient Response Method
.... H. Kobayashi and M. Kobayashi   117
Discussion   140

Application of Computers to Kinetic Studies
    V. B. Skomorokhov, M. G. Slin'ko and V. I. Timoshenko 149
Improved Steady-Rate Treatment of a Chemical Reaction with
  Regard to Selectivity of Complex Catalysis
    ....K. Miyahara 163
On the Problem of Classifying Catalytic Reactions
    .... V. A. Roiter and C. I. Golodets 185
Studies of Heterogeneous Catalysis by the Pulse Reaction
  Technique
    .... Y. Murakami 203
Discussion 218

Activation of Molecular Nitrogen and Saturated Hydrocarbons by
  Transition Metal Complexes
    ....A. E. Shilov 231
Oxidation of Olefins by Mercuric Ion
    .... Y. Saito 241
Optimization of Catalysts for Hydrogenation in Solutions
    .... D. V. Sokol'sky and A. M. Sokol'skaya 259
Overvoltage and the Kinetic Law of Elementary Stages of Reaction
  of Hydrogen Platinum and Nickel Cathode Deposit in Alkaline
  Solutions
    .... A. Matsuda, R. Notoya and T. Ohmori 273
On the Mechanisms of Enzymatic Reactions
    ....O. M. Poltorak and E. S. Chukharai 295
Discussion 321

SOME THEORETICAL PROBLEMS OF CATALYSIS

Introductory Remarks

The gap between the performance of investigations in catalysis and the possibility of introducing the results in scientific journals continuously increases reaching now 1.5 - 2.5 years. Therefore, new ways of mutual information exchange among scientists are necessary to satisfy the modern rate of development of the theory of catalysis and its practical use. For researchers of different countries, small international seminars for discussing current problems in various fields of catalysis can be one of the possible forms. That is why the Scientific Council on Catalysis of the USSR Academy of Sciences and the Catalytic Society of Japan decided to hold annual conferences on catalysis in each country in turn.

The First Soviet-Japanese Seminar was held from July 4 to 8, 1971, in Novosibirsk at the Institute of Catalysis of the USSR Academy of Sciences. For us, the scientists of Novosibirsk, this seminar was significant in many respects. There is no doubt about the rapidly increasing importance of catalysis for scientific-technical progress: The advances of chemical, oil-refining, and other fields of industry are intimately associated with the development of catalytic methods. Catalysis is sure to play a leading role in rapid development of Siberian industry in using its extremely rich natural resources, in particular, the recently discovered unique oil and gas deposits in West Siberia.

It is very important that in the discussion of problems of catalysis prominent Japanese scientists took part who have achieved considerable advances both in fundamental researches and in applying catalytic methods in the chemical industry developing extremely rapidly in Japan. It is also important that in the work of the seminar, along with the scientists of the Institute of Catalysis, the most prominent specialists of the USSR from Moscow, Kiev, Alma-Ata, and other cities took part, providing a high scientific level and activity in discussing report and communications.

The very broad subject chosen-"Theoretical Problems of Catalysis"-was quite reasonable for the first meeting. A total of 18 reports and 11 communications were made; they gave rise to an animated discussion with the participation of more than 60 scientists. The problems of heterogeneous, homogeneous, and fermentative catalysis were taken up in the reports, including the problems of the theory of choice of catalyst, the mechanism, kinetics, and methods of investigation by the examples of oxidation, hydration, electrochemical catalysis, isotopic exchange, and other reactions.

In spite of such variety in the work performed in different countries and cities spaced more than 10,000 km apart, some common tendencies are clearly seen. In most papers presented the tendency is to a profound understanding of the catalytic reaction mechanism including the elementary stages of interaction of reactants with a catalyst. Reasonable attention was given to the change of electron filling of catalyst and reactant orbitals, the influence of ligands or adjacent atoms of a catalyst on it, the possibility of electron transfers with the help of the system of conjugated bonds, etc.

The latter was clearly brought out in the report by the head of the Japanese delegation Professor T. Kwan (Tokyo), concerning the mechanism of interaction with the participation of transition metal ions performed through the conjugated bonds of a porphyrin ring. The same trend for profound understanding of the mechanism of interaction of catalysts in homogeneous catalysis by transition metal complexes is characteristic of the paper by Professor A. E. Shylov (Moscow) dedicated to the reactions with the participation of molecular nitrogen and saturated hydrocarbons. The report of Professor K. Tamaru (Tokyo) on the catalytic action of donor-acceptor complexes attracted considerable interest as well as his communication on significant change of electron structure of iron catalysts by using alkaline additives.

The rules of complex formation are also reasonably important in the process of heterogeneous catalysis. It was shown in the report of Professor O. V. Krylov (Moscow) who investigated the formation of intermediate complexes and coordination of transition metal ions in them by physical methods in reactions of catalytic oxidation on oxide catalysts. The same problem was discussed in the report of Professor V. B. Kazanski et al. (Moscow), who investigated the formation of intermediate complexes in heterogeneous catalysis and their participation in elementary stages of the catalytic reaction cycle by optical and EPR methods.

It is of particular interest that features similar to those of homogeneous and heterogeneous catalysis also appear in fermentative catalysis as was shown in the report of Professor O.M. Poltorak (Moscow). Revealing common tendencies in such different fields of catalysis seems to be the advantage of organizing seminars with broad subjects.

Common problems of classifications of catalytic reactions and forecasting of catalytic action were discussed in the reports of Professors V.A. Roiter, G.I. Golodets (Kiev), and the author.

A reasonable number of reports cover problems of the kinetics of heterogeneous catalysis reactions and experimental methods of determining catalytic transformation rates. The report of Professor M.I. Temkin (Moscow) gives an original conception explaining variations of heat and energy of activation of surface reactions with filling. Professor K. Miyahara (Sapporo) developed the theory of complex multiroute reactions of heterogeneous catalysis based on the concepts by Professor J. Horiuti. In the report of Professor M.G. Slinko et al. (Novosibirsk), the possibility of using computers (EVMs) in kinetic investigations was considered. The advantages of on-line connection of experimental systems with a computer were shown, and methods of deducing kinetic equations by the Chebyshev method of smoothing were considered. The possibilities of a pulse method for studying the kinetics and mechanism of catalytic reactions was discussed in detail in the report of Professor Y. Murakami (Nagoya). In the interesting investigation of Professors H. and M. Kobayashi (Sapporo), a response method for investigating the mechanism of CO oxidation by manganese dioxide was fruitfully used.

From separate groups of catalytic processes, oxidation reactions were most completely represented in the seminar. This reflects the rapidly growing practical importance of these reactions, in particular, of selective oxidation to produce the most important chemical products, as well as obvious advances in developing the theory of these reactions.

The rules of catalytic oxidation were considered in many of the above-mentioned reports. In the report of Professor I. Moro-oka (Tokyo) it was shown that the processes of selective oxidation in heterogeneous catalysis could proceed in a more complicated way than it was thought earlier, in particular, in oxidation the water oxygen can play a decided role. Professor T. Seiyama (Fukuoka) gave the results of investigating numerous oxide catalysts of propylene partial oxidation and made an attempt to classify them on the basis of semiconductive and other properties. The mechanism of

reactions of selective oxidation were discussed in the communication of Professor Y. B. Gorokhvatski (Kiev).

The results of investigating the mechanism of oxidation of
CO and hydrogen and isotopic exchange were given in four
communications by members of the Institute of Catalysis
(V. V. Popovski, E. A. Mamedov, V. I. Marshneva,
V. D. Sokolovski, L. A. Sazanov, E. V. Artamonov,
V. S. Muzykantov, G. I. Panov, and the author). The report
of Professor Y. Saito (Tokyo) is dedicated to the mechanism
of homogeneous catalytic oxidation of olefins by transition
metal complexes. The role of oxide catalysts in the process
of liquid-phase chain oxidation of isopropylbenzene was
considered in the communication of E. A. Blumberg et al.
(Moscow).

The catalytic properties of clean metal surfaces obtained
in a superhigh vacuum were discussed in the communications
of Professor M. I. Tretyakov (Moscow) and V. I. Savtchenko
and the author (Novosibirsk). The rules of electrochemical
catalysis were considered in the report by Professor
A. Matsuda, R. Notoya and T. Ohmori (Sapporo). The
report of Professor D. V. Sokolski and A. M. Sokoloskaya
(Alma-Ata) was dedicated to electrochemical methods of
investigating metallic catalysts acting in a liquid phase and
the mechanism of hydrogenation reactions.

Some communications were dedicated to catalytic decomposition of hydrogen peroxide (A. P. Purmal, Moscow),
catalytic polymerization (Yu. I. Ermakov et al., Novosibirsk)
and catalytic transformations of sulfur-organic compounds
(A. V. Mashkina et al., Novosibirsk).

In conclusion I would like to express my hope that the
first meeting in Novosibirsk will be the beginning of regular
work of the Soviet-Japanese Seminar on Catalysis and
serve as further consolidation of the scientific relations
between the scientists of our countries.

G. K. Boreskov

Mechanism of the Selective Oxidation of Olefin by Metal Oxides

Y. Moro-oka and Y. Takita

Tokyo Institute of Technology, Tokyo

INTRODUCTION

It has generally been found in oxidation reactions that the oxygen atom is incorporated into the oxidized product in two different ways. One is the case where the oxygen atom is incorporated from molecular oxygen; the second is where the atom is incorporated from a water molecule

$$A + \tfrac{1}{2}O_2 \longrightarrow AO \tag{1}$$
$$A + \tfrac{1}{2}O_2 + H_2O^* \longrightarrow AO^* + H_2O \tag{2}$$

where A is the molecule to be oxidized. For example, in biochemical oxidation, Eq. (2) is commonly observed and is catalyzed by oxidase, while it has been proved that the oxidation catalyzed by oxygenase[1] or mixed function oxidase[2] follows Eq. (1).

In recent years, a number of catalytic oxidations of hydrocarbons have been developed using the transition metal oxide catalysts[3]. The mechanisms of these oxidations have been studied extensively by the tracer technique with regard to the behavior of the hydrocarbon[4-8]. However, little work has been done on the behavior of active oxygen. No reaction of the type represented by Eq. (2) has ever been reported in heterogeneous oxidation on transition metal oxide catalysts.

The present investigation was started in an attempt to discriminate between Eqs. (1) and (2) in the heterogeneous oxidation of olefins. A clear discrimination of the route of oxygen incorporation into the oxidized product will undoubtedly be helpful for understanding both the mechanism of the oxidation and the role of the active site on the oxide surface. In this work, oxidation of propylene over SnO_2-MoO_3, MoO_2-Bi_2O_3, and Pd-activated charcoal catalysts was studied by means of $H_2{}^{18}O$ tracer techniques. It was found that both types of oxygen incorporation occur depending on the type of product or catalyst. The mechanism and selectivity of the heterogeneous oxidation of olefins will be discussed from the standpoint of oxygen incorporation.

EXPERIMENTAL SECTION

Catalysts.   Two types of SnO_2-MoO_3, (A) and (B), were used as catalysts. The composition of both catalysts was Sn/Mo = 9/1 in metal atom ratio. SnO_2-MoO_3 (A) was prepared by the decomposition of stannous hydroxide mixed with ammonium molybdate solution. The oxide powder was pressed into cylindrical pellets and calcined in air at 550°C for 5 hr. These pellets were crushed and sieved to 28-35 mesh. The surface area of the sample measured by the B.E.T. method was 45.6 m^2/g. SnO_2-MoO_3 (B) was prepared from ammonium molybdate and stannic oxide which was obtained by the calcination of stannous hydroxide in air at 800°C for 5 hr. The stannic oxide (24-35 mesh) was impregnated with ammonium molybdate solution, dried on a water bath, and then calcined in air at 550°C for 5 hr. The B.E.T. surface area of the sample was 7.4 m^2/g.

MoO$_2$-Bi$_2$O$_3$ was prepared from ammonium molybdate and bismuth nitrate. The composition was Mo/Bi = 4/3 in metal atom ratio and the surface area was 2.9 m^2/g. Pd-activated charcoal catalyst was prepared from palladium chloride and 10-20 mesh activated charcoal. Palladium chloride supported on activated charcoal was reduced to metallic palladium by hydrogen at 250°C for 5 hr and was activated by exposure to the reactant gas stream for 10 hr at the reaction conditions before the run.

Procedure.   All the oxidation runs were carried out using a flow system at 1 atm pressure. A Pyrex tube reactor of 8 mm diameter was heated by a tubular electric heater. The temperature of the catalyst bed was controlled by means of a thermo-electric controller. The gaseous reactants were introduced from cylinders through needle valves, and water was introduced by evaporation into the gaseous reactant stream. The reaction products were cooled by water, separated into gaseous or liquid phases, and analyzed by gas chromatography and mass spectrometry. The gc columns for separation and analysis of the products were Molecular Sieve 5A for oxygen and carbon monoxide; β,β'-oxodipropionitrile supported on 60-80 mesh Al_2O_3 for carbon dioxide and propylene; polyethylene glycol supported on 60-80 mesh Celite for acetone, acrolein, and acetaldehyde; and dioctyl sebacate with 5% sebacic acid supported on 60-80 mesh Celite for acetic acid, propionic acid, and acrylic acid. Determination of ^{18}O content of the reaction products.   The ^{18}O content of the oxidized products was determined by mass spectrometry after separation and purification by gc.   The mass spectra obtained at 80 ionization voltages were corrected for natural isotopes, and the ionization efficiency for the isomer containing ^{18}O was assumed equal.   The ^{18}O content of the oxidized products was calculated from the peak height ratio in the corrected mass spectra.

The ^{18}O content of the unreacted oxygen was determined by mass spectrometry after it was converted to carbon dioxide. The unreacted oxygen gas was separated by a Molecular Sieve 5A column and then introduced into a closed circulating reaction system where it was completely converted to carbon dioxide by reaction with a stoichiometric amount of carbon monoxide, using Pt-asbestos catalyst at 250°C.

RESULTS AND DISCUSSION

Route of Oxygen Incorporation
1. Oxidation of propylene to acetone over SnO_2-MoO_3

The author reported previously that propylene is oxidized selectively to acetone on stannic oxide combined with molybdenum trioxide at 100-200°C.[9]   The reaction was also observed on several transition metal oxides combined with molybdenum trioxide.[10]   It was decided to extend the study of the mechanism of propylene oxidation on the SnO_2-MoO_3 catalyst because the main product in this reaction was quite different from that reported for the usual heterogeneous oxidations of olefin on transition metal oxide catalysts.

Propylene was oxidized to acetone in the presence of water vapor enriched with ^{18}O on the SnO_2-MoO_3 (A) catalyst under two different conditions.   The oxidation to acetone was quite selective, as reported previously.   The ^{18}O content of acetone, unreacted oxygen and water is summarized in Table I.   In every run, the ^{18}O content of each of the products was determined, rejecting those obtained in the first 1 hr.   As can be seen in Table I, ^{18}O tracer was detected in the acetone produced.   Although the values of ^{18}O content in the acetone produced are slightly smaller than those of the water in the reaction system, it is obvious that the oxygen atom from the water molecule is incorporated into the oxidized

Table I. ^{18}O Content of Products in the Oxidation of Propylene to Acetone.[a]

Run	Reaction temperature (°C)	GHSV $\left(\dfrac{\text{ml-STP}}{\text{ml-cat.hr}}\right)$	C_3 conversion (%)	Selectivity to acetone (%)	^{18}O content (atomic %)			
					H_2O (input)	H_2O (output)	CH_3COCH_3	O_2 (output)
1[b]	125	600	3.99	93.7	5.48	4.62	4.11	0.02
2[b]	125	600	3.99	93.7	5.48	4.91	3.68	0.04
3[b]	125	600	3.99	93.7	5.48	4.59	3.93	0.01
4[b]	125	600	3.99	93.7	5.48	4.78	4.55	0.02
5[b]	125	600	3.99	93.7	4.59	–	2.72	tr
6[c]	225	2000	1.67	68.1	8.60	8.47	6.73	0.01

a   Reactant gas composition: C_3H_6 20 vol%, O_2 30%, H_2O 30%, N_2 20%.

b   Flow rate of total reactant gas: 21.8 ml-STP/min.
    Catalyst: SnO_2-MoO_3 (A) 3.6 g.

c   Flow rate of total reactant gas: 34.3 ml-STP/min.
    Catalyst: SnO_2-MoO_3 (A) 2.0 g.

product.

Two oxygen exchange reactions were examined in connection with the oxygen incorporation. One is the oxygen exchange reaction between molecular oxygen and water:

$$^{16}O_2 + H_2{}^{18}O \rightleftharpoons {}^{16}O\,{}^{18}O + H_2{}^{16}O \qquad (3)$$

The ^{18}O content of the unreacted oxygen is listed in the last column in Table I. The oxygen exchange reaction between molecular oxygen and water was scarcely observable under the conditions adopted in these runs. The result eliminates the possibility that ^{18}O in the acetone was incorporated from $^{16}O\,{}^{18}O$ molecules formed by the exchange reaction expressed in Eq. (3).

Another oxygen exchange reaction between acetone and water, shown in Eq. (4),

$$CH_3\text{-}\underset{\underset{16}{\overset{\|}{O}}}{C}\text{-}CH_3 + H_2{}^{18}O \rightleftharpoons CH_3\text{-}\underset{\underset{18}{\overset{\|}{O}}}{C}\text{-}CH_3 + H_2{}^{16}O \qquad (4)$$

was also checked using water enriched with ^{18}O. The contact time of acetone and its concentration in the reactant gas were adjusted to the same levels as those produced in the oxidation runs. The reactant gas composition and the other conditions were also the same as adopted in the oxidation runs, except that propylene in the reactant gas was replaced by nitrogen. The results are shown in Table II.

Although some exchange reaction was observed under these conditions, the rate was slower than that of the oxidation. The ^{18}O content observed in the acetone produced is not attributable to the exchange reaction. Thus the possibility that ^{18}O is incorporated into acetone by an exchange reaction between water and acetone which is formed by a reaction between propylene and molecular oxygen can be rejected.

Propylene is oxidized to acrolein and acetaldehyde as well as to acetone on $SnO_2\text{-}MoO_3$ at higher reaction temperature.[11] This oxidation was also investigated using $H_2{}^{18}O$ tracer in a similar manner. A low surface area catalyst, $SnO_2\text{-}MoO_3$ (B), was used for this oxidation. The results are summarized in Table III.

Table II. Oxygen Exchange Reaction between Acetone and Water on the SnO_2-MoO_3 (A) Catalyst.[a]

Run	^{18}O content (%)	
	H_2O (input)	CH_3COCH_3
1	5.48	2.36
2	5.48	2.65
3	4.59	1.29
4	4.59	1.69

a Reaction temperature: 125°C.
Reactant gas composition: oxygen 40.9 vol%, nitrogen 33.7%, water 24.6%, acetone 0.87%.
GHSV: 600 ml-STP/ml-cat. hr.

Table III. ^{18}O Content of Products in the Oxidation of Propylene to Form Acetone, Acrolein, and Acetaldehyde.

Reaction temperature (°C)	^{18}O content (atomic %)			
	CH_3COCH_3	CH_2=CHCHO	CH_3CHO	O_2 (output)
365	2.21	0.20	0.43	0.54
366	3.51	0.11	0.43	0.19
367	2.07	0.34	0.48	0.20
370	2.38	0.00	0.45	0.58
366	2.62	0.03	0.47	0.32
average	2.56	0.14	0.45	0.37

Note ^{18}O content of water: input 6.70, output 5.5%.
Reactant gas composition is identical with that in Table I.
Flow rate of total reactant gas: 37.6 ml-STP/min.
Catalyst: SnO_2-MoO_3 (B) 2.0 g.
GHSV: 2260 ml-STP/ml-cat. hr.

In the oxidation at 370°C, ^{18}O was again incorporated into the acetone. * In contrast with the oxidation to acetone, ^{18}O tracer was scarcely found in the acrolein produced. A small amount of ^{18}O was found in the acetaldehyde. Since acetaldehyde is formed by the oxidation of acetone as well as by the direct oxidation of propylene under these conditions, this incorporation of ^{18}O in the acetaldehyde may be due to the consecutive oxidation of acetone.

As described above, it was found that the labeled oxygen atom in water was incorporated into the acetone. The incorporation does not depend on an exchange reaction between molecular oxygen and water or between the produced acetone and water (Tables I and II), but depends closely on the mechanism of the oxidation to form acetone. Two probable mechanisms may be considered for the incorporation of the oxygen atom. Since it is generally accepted that the active species for the oxidation derives from molecular oxygen and is an atomic one bearing some negative charge, the possible mechanisms may be sketched as shown in Fig. 1. One mechanism shows that acetone is produced by a reaction of propylene (or some derivative of propylene) with the active species derived from molecular oxygen, which is in rapid equilibrium with the water on the oxide surface. When the oxygen exchange reaction between gaseous oxygen and the adsorbed species is not rapid, this rapid oxygen equilibration on the catalyst surface cannot be detected. The second mechanism is that acetone is produced by a direct interaction between the active species of propylene and water. It seems to be difficult to discriminate between these two mechanisms directly. However, when oxidation proceeds according to mechanism I, ^{18}O must be incorporated into all of the oxidation products which are formed simultaneously on the catalyst surface. In practice, as shown in Table III, ^{18}O was only

---

* The ^{18}O content of the acetone produced is considerably lower than that of the water. The values were confirmed by several further runs using $H_2{}^{18}O$. Since a significant amount of water is formed at these reaction conditions, dilution of the ^{18}O content of water on the catalyst surface by water produced in the oxidation would be more marked than in the gaseous state. Observed low values of ^{18}O content of the acetone seem to be due to this dilution on the catalyst surface.

Mechanism I

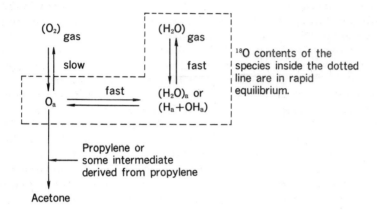

Mechanism II

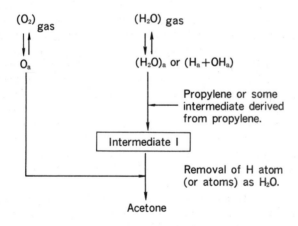

Fig. 1

found in acetone. It was scarcely present in acrolein, which was formed simultaneously with acetone on the catalyst surface. This suggests clearly that the rapid oxygen exchange reaction depicted in mechanism I does not occur on the oxide surface. Thus, it can reasonably be concluded that ^{18}O is incorporated into acetone by direct interaction between the active species of water and propylene (or some derivative of propylene) as depicted in mechanism II.

2. Oxidation of propylene to acrolein on different catalysts

Oxidation of propylene to form acrolein has been reported over a number of transition metal or transition metal oxide catalysts. Further investigation was done on the incorporation of the oxygen atom into acrolein over two different catalysts.

First MoO_3-Bi_2O_3, which is known as a famous allylic oxidation catalyst, was examined. The results are summarized in Table IV. As can be seen in Table IV, the ^{18}O content of acrolein is far lower than that of the water in the reactant gas. It can be concluded that the oxygen atom in acrolein comes from molecular oxygen (in some cases via bulk oxide), not from the water molecule, although it is well known that presence of water vapor in the reactant gas mixture increases the selectivity of conversion to acrolein.

Oxygen incorporation was also examined on a Pd-activated charcoal catalyst. This oxidation was recently reported by Fujimoto et al.[12] and by Yamazoe et al.[13] independently. The results, using H_2 ^{18}O tracer, are presented in Table V.

In contrast to the results on the SnO_2-MoO_3 and MoO_3-Bi_2O_3 catalysts, ^{18}O was found in the acrolein produced on the Pd-activated charcoal catalyst. Although some oxygen exchange between molecular oxygen and water was observed on the catalyst, the rate was too slow to account for the ^{18}O content in the acrolein. The ^{18}O content of acrolein cannot be explained by an exchange reaction between the produced acrolein and water. The oxygen atom in acrolein is evidently incorporated from water, not from molecular oxygen, on the Pd-activated charcoal catalyst. Thus different ways of oxygen incorporation are realized, depending upon the type of catalyst, in the oxidation of propylene to form the same oxidation product.

Mechanism of the Oxidation

1. Oxidation to acetone

The fact that the incorporation of oxygen into acetone

Table IV. $H_2^{18}O$ Tracer in the Oxidation of Propylene to
Acrolein over MoO_3-Bi_2O_3 Catalyst[a]

Run	Reaction temperature (°C)	C_3 conv. (%)	Selectivity to acrolein (%)	^{18}O content (atomic %)	
				H_2O^b	CH_2=CHCHO
1	360	2.50	73.0	3.92	0.26 ± 0.2
2	385	4.35	65.5	3.84	0.00 ± 0.2
3	433	9.45	55.0	3.73	0.16 ± 0.2
4	460	14.80	51.0	3.40	0.18 ± 0.2

a   Gas composition is same as in Table I.
GHSV: 600 ml-STP/ml-cat. hr.
b   Average of the value for input (4.00) and output gas.

Table V. $H_2^{18}O$ Tracer in the Oxidation of Propylene to Acrolein
over Pd-activated Charcoal Catalyst[a]

Type of reaction	Reaction temperature (°C)	C_3 conv. (%)	Selectivity to acrolein[d] (%)	^{18}O content (atomic %)	
				CH_2=CHCHO	O_2 (output)
Oxidation[b]	125	3.6	56	5.09	-
	125	3.6	56	5.10	1.90
Exchange[c]	130	-	-	0.75	-
	130	-	-	1.13	-

a ^{18}O content of input water: 5.48%.
b   Flow rate of reactant: oxygen 8.0, propylene 10.0, water 27.6
ml-STP/min.
GHSV: 530 ml-STP/ml-cat. hr.
Catalyst: 1.9 g.
c   Exchange reaction between acrolein and water. Conditions are
the same as for the oxidation runs except that propylene in the
reactant gas is replaced by nitrogen.
d   Other products: acrylic acid 21%, carbon dioxide 19%.

follows Eq. (2), suggests some possible mechanisms for
acetone formation (Fig. 2). One is analogous to the mechanism for metal ion oxidation in aqueous solution.[14-16]   A
π-complex formation between olefin and metal ion is followed

by insertion of an OH group into the π-bond to form a σ-complex, which decomposes to a carbonyl compound with displacement of the hydrogen atom. Another mechanism involves an intermediate such as CH_3CHCH_3, which is probably a carbonium ion. The attack of an OH group or water on this intermediate forms a secondary alcohol which is easily oxidized to a carbonyl compound.

If the oxidation obeys mechanism II (Fig. 2) it is expected that the surface acidity will play an important role in the propylene oxidation. Accordingly, the oxidation was further investigated with regard to its dependence on the acidic nature of the oxide catalysts. The amount of acidic sites stronger than $pK_a = 3.3$ was determined by n-butylamine adsorption using dimethyl yellow as indicator. Catalytic activity for double bond isomerization of 1-butene was also examined. These acidic properties of the catalysts were compared with the activities for propylene oxidation in Table VI.

As can be seen in Table VI, the larger the amount of acidic sites, the higher the activity for propylene oxidation. The activity for 1-butene isomerization also corresponds to that for propylene oxidation. In addition, a hydrogen exchange reaction between propylene and water was observed on SnO_2-MoO_3, involving an intermediate such as CH_3CHCH_3.[17] Furthermore, a small amount of secondary alcohol was detected in the oxidation runs.[10] On the other

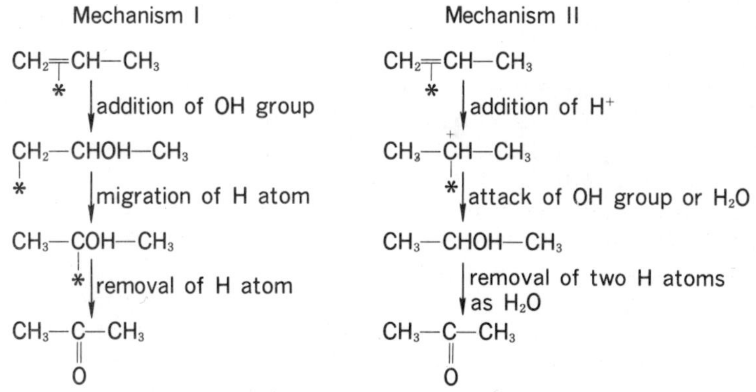

Fig. 2

Table VI.   Relation between Catalytic Activity for
            Propylene Oxidation and Acidic Properties of
            the Catalysts.

Atomic ratio in SnO₂-MoO₃ Sn : Mo	Catalytic activity[a]			Amount of acidic sites (10^{18} /m²-cat)
	C_3H_6[b] oxid.	1-C_4H_8 isomer.	C_3H_8 oxid.	
100 : 0	1.70	1.81	1.44	0.007
99 : 1	1.77	1.80	1.43	0.46
98 : 2	1.89	2.29	1.53	0.48
95 : 5	1.91	2.12	1.53	0.63
90 : 10	2.13	2.55	1.63	2.50
50 : 50	2.19	2.57	1.64	2.94
5 : 95	2.06	2.28	1.59	1.36

a   Reciprocal temperature ($°K^{-1}$) at which the reaction rate
    reachs $5 \times 10^{-3} \mu$mole/m²-cat. sec.
b   Selectivity to acetone is about 70% of converted propylene
    except in the cases of pure SnO_2 and Sn/Mo = 99/1
    catalysts.

hand the catalytic activity for propane oxidation, which
depends exclusively on the reactivity of active species derived from molecular oxygen,[25]   does not change with the
catalyst composition.   On the basis of these results, mechanism II (Fig. 2) seems to be more reasonable for acetone
formation.

2. Oxidation to acrolein

    It has been well established that propylene is oxidized to
acrolein via an allylic intermediate on MoO_3-Bi_2O_3 catalyst.[5-7]
A mechanism which also involves a π-allyl intermediate was
proposed for oxidation on the Pd-activated charcoal
catalyst,[12]   although concrete evidence for the mechanism
has not been obtained.   The result that the oxygen atom is
incorporated into acrolein from water strongly supports the
proposed mechanism.   Since both terminal carbon atoms in
the allyl radical have a partial positive charge, it seems
reasonable that the OH group attacks either of the terminal
carbon atoms to form acrolein.   Simplified mechanisms for
the oxidation on both catalysts are shown in Fig. 3.

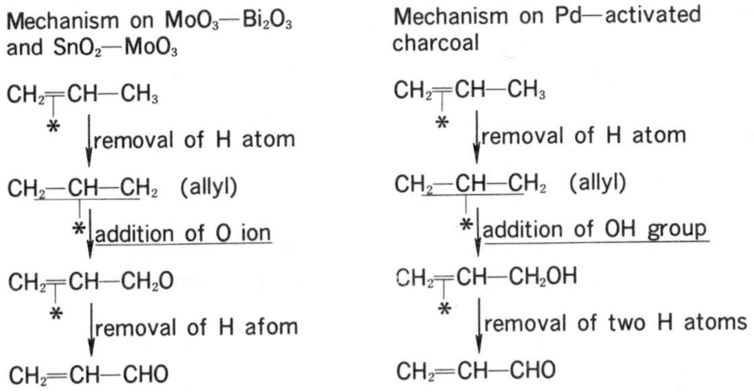

Mechanism on MoO₃—Bi₂O₃ and SnO₂—MoO₃

Mechanism on Pd—activated charcoal

$CH_2\!=\!CH\!-\!CH_3$
*
↓ removal of H atom

$CH_2\!-\!CH\!-\!CH_2$ (allyl)
* ↓ addition of O ion

$CH_2\!=\!CH\!-\!CH_2O$
*
↓ removal of H afom

$CH_2\!=\!CH\!-\!CHO$

$CH_2\!=\!CH\!-\!CH_3$
*
↓ removal of H atom

$CH_2\!-\!CH\!-\!CH_2$ (allyl)
* ↓ addition of OH group

$CH_2\!=\!CH\!-\!CH_2OH$
*
↓ removal of two H atoms

$CH_2\!=\!CH\!-\!CHO$

Fig. 3

Selectivity of Olefin Oxidation

In the course of the study of heterogeneous oxidation by transition metal oxide catalysts, the reaction mechanism has generally been considered as involving only molecular oxygen or oxygen in the bulk oxide. However, the results obtained in this work present definite evidence that the reaction involves water, depending on the type of product or catalyst. The selectivity found in the oxidation of olefins must be reconsidered in view of this fact. First, let us consider the product distribution at different reaction temperature.

Product distributions found in the oxidation of propylene over SnO_2-MoO_3 (A) and (B) are shown in Figs. 4 and 5. At lower reaction temperatures, acetone is the main product. On the other hand, acrolein and acrylic acid are mainly formed at higher temperature. It can be seen that the source of the oxygen atom which is incorporated into oxidized products changes from water to molecular oxygen with increasing reaction temperature. The product distributions shown in Figs. 4 and 5 have been observed not only on the SnO_2-MoO_3 catalysts but also on a number of transition metal or metal oxide catalysts.[10]

The above product distribution is understandable on the basis of different ways of oxygen incorporation. Since the two independent oxygen incorporations are competitive, the selectivity of the oxidation is determined by their relative

rates. It has been pointed out that the weaker the adsorption
of oxygen, the higher the catalytic activity of the metal
oxide.[18−22]   This suggests that the rate of the reaction
depending on the active species derived from molecular oxygen increases monotonically with the reaction temperature.
On the other hand, hydration of the olefin, involving an OH
group, becomes unfavourable with rising reaction temperature as can be seen from the equilibrium constant for
isopropyl alcohol formation from propylene and water. Thus,
oxyhydration to form acetone is predominant at lower temperature, while oxidation to form acrolein is mainly observed
at higher temperature.

The reaction products expected from the interaction
between the active species of oxygen and olefin are summarized in Table VII for propylene and 1-butene. It has been
generally accepted that oxygen is adsorbed onto metal or
metal oxide as an electron acceptor and forms O^- or O^{2-}
species. On the other hand, electron donating adsorption of
olefin was observed both in static adsorption[23] and in adsorption during the oxidation reaction.[24, 25]   Since dissociatively
adsorbed oxygen accepts electrons from the bulk oxide and
loses its electrophilic nature, no electrophilic addition
reaction to form epoxide or saturated carbonyl compounds

Table VII.   Reaction Products Expected from the Combination
between the Active Species of Olefin and Oxygen.

Source of oxygen (active form)	Active species of olefin	Product		Typical catalyst
		Propylene	1-Butene	
O_2 (O^- or O^{2-})	π-complex	Allyl[a]	Allyl[b]	MoO_3-Bi_2O_3
	Allyl	Acrolein	Butadiene	MoO_3-Bi_2O_3
	Carbonium ion	-	-	-
H_2O (OH or OH$^-$)	π-complex	Acetone	M. E. K.[c]	$PdCl_2$-$CuCl_2$
	Allyl	Acrolein	M. V. K.[d]	Pd-A. C.
	Carbonium ion	Acetone	M. E. K.	MoO_3-SnO_2

a CH_2-CH-CH_2.
b CH_2-CH-CH-CH_3.
c   Methyl ethyl ketone.
d   Methyl vinyl ketone.

can be expected in the reaction involving active species derived from molecular oxygen. Actually, oxidative dehydrogenation to form an allylic intermediate has mainly been reported in the heterogeneous oxidation of olefin at higher temperatures. On the other hand, saturated carbonyl compounds such as acetone and methyl ethyl ketone are the main products in metal ion oxidation in aqueous solution.[14-16] Oxygen incorporation from water has been commonly observed in metal ion oxidation. There has been marked difference between these two oxidation, i. e., oxidation by metal oxide catalysts and that by metal ions. The group of oxyhydration reactions which is reported in this work, and others which may be discovered in the future, shows that these reactions are not entirely dissimilar.

References

1) O. Hayaishi, M. Katagiri, and S. Rothberg, J. Am. Chem. Soc., 77, 5450 (1955).
2) H. S. Mason, W. L. Fowlks, and E. Peterson, J. Am. Chem. Soc., 77, 2915 (1955).
3) R. J. Sampson and D. Shooter, in Oxidation and Combustion Reviews (C. F. H. Tipper, ed.), Vol. 1, p. 223, Elsevier, Amsterdam, 1965; L. Ya. Margolis, Advan. Catalysis, 14, 429 (1963).
4) H. H. Voge, C. D. Wagner, and D. P. Stevenson, J. Catalysis, 2, 58 (1963).
5) C. R. Adams and T. J. Jennings, J. Catalysis, 2, 63 (1963); ibid., 3, 549 (1964).
6) C. C. McCain, G. Gough, and G. W. Godin, Nature, 198, 989 (1963).
7) W. M. H. Sachtler, Rec. Trav. Chim., 82, 243 (1963).
8) F. L. J. Sixma, E. F. J. Duynstee, and J. L. J. P. Hennekens, Rec. Trav. Chim., 82, 901 (1963).
9) Y. Moro-oka, S. Tan, Y. Takita, and A. Ozaki, Bull. Chem. Soc. Japan, 41, 2820 (1968); S. Tan, Y. Moro-oka, and A. Ozaki, J. Catalysis, 17, 132 (1970).
10) Y. Moro-oka, Y. Takita, and A. Ozaki, J. Catalysis, in press.
11) J. Buiten, J. Catalysis, 10, 188 (1968).
12) K. Fujimoto, H. Yoshino, and T. Kunugi, in Preprint

for 3rd Oxidation Symposium (Tokyo), p. 111 (1970).

13) N. Yamazoe, M. Aramaki, J. Hojyo, and T. Seiyama, in Symposium of Catalysis Soc. Japan (Nagoya), 1970.

14) P. M. Henry, J. Am. Chem. Soc., 86, 3246 (1964); ibid., 88, 1595 (1966).

15) Y. Saito, in this volume, P. 202.

16) P. M. Henry, J. Am. Chem. Soc., 87, 990, 4423 (1965).

17) J. Buiten, J. Catalysis, 13, 373 (1969).

18) A. P. Dzisjak, G. K. Boreskov, L. A. Kasatkina, and V. K. Kochurihin, Kinetika i Kataliz, 2, 386, 727 (1961); ibid., 3, 81 (1962); G. K. Boreskov, Advan. Catalysis, 15, 285 (1964).

19) V. V. Popovsky and G. K. Boreskov, Probl. Kinetika i Kataliza, Akad. Nauk. SSSR., 10, 67 (1960); I. Komuro, H. Yamamoto, and T. Kwan, Bull. Chem. Soc. Japan, 36, 1532 (1964).

20) Y. Moro-oka and A. Ozaki, J. Catalysis, 5, 116 (1966).

21) Y. Moro-oka, Y. Morikawa, and A. Ozaki, J. Catalysis, 7, 23 (1967).

22) G. K. Boreskov, V. V. Popovsky, and V. A. Sazonov, in 4th Int. Congr. Catalysis (Moscow), 1968, No. 33.

23) E. Kh. Enikeev, L. Ya. Margolis, and S. Z. Roginskii, Dokl. Akad. Nauk. SSSR, 124, 606 (1959); E. Kh. Enikeev, O. V. Isaev, and L. Ya. Margolis, Kinetika i Kataliz, 1, 431 (1960).

24) Y. Moro-oka and A. Ozaki, J. Am. Chem. Soc., 89, 5124, (1967).

25) Y. Moro-oka, M. Otsuka, and A. Ozaki, Trans. Farad. Soc., 67, 877 (1971).

Optimal Bond Energies with Catalyst of Reactants and Catalytic Reaction Products

G. K. BORESKOV

Institute of Catalysis, Siberian Branch, USSR Academy of Sciences, Novosibirsk

In catalytic reactions the catalyst is a component of the activated complexes at most stages of reaction. To predict catalytic activity, i. e., catalytic reaction rates, it is neces sary to know energies of formation for these activated complexes for all stages of the reaction pathway. Unfortunately, quantum chemical calculations of these energies cannot be realized yet, even for the simplest chemical interactions. However, an approximate estimation is possible of the energy variation of active complexes in a variety of catalysts in relation to a certain reaction or for catalytic conversion of various substances on a particular catalyst. It is based on the Brönsted-Polani-Semenov (BPS) linearity rule for activation energy (E) and heat effect of chemical reaction (q):

$$E = E_0 \pm \alpha q \tag{1}$$

Here E_0 and α are constant coefficients, α being in the range of O - 1.

The physical nature of this relation in the case of catalytic reactions amounts to the fact that in the active complex at each stage the energy of the chemical bond with the catalyst forms a particular part, α, of the energy of the corresponding bonds forming or breaking at this reaction stage. According to this, the plus sign in Eq. (1) corresponds to the bonds breaking at this stage, and the minus sign to those forming.

From the BPS relation the existence of optimal values of reagent-catalyst bond energies follows.

Thus, for the simplest two-stage scheme of reaction

$$A + B \longrightarrow \Sigma P_i,$$

passing through the stages

$$A + K \longrightarrow AK \tag{I}$$

23

$$\text{and} \quad AK + B \longrightarrow K + \Sigma P_i \qquad\qquad (\text{II})$$

the activation energies will depend on the heat of intermediate interaction of the reactant A with the catalyst K in the following way:

$$E_I = E_{I_0} - \alpha_I q \qquad\qquad (2)$$

$$E_{II} = E_{II_0} + (1 - \alpha_{II}) q \qquad\qquad (3)$$

According to this, the stage rates are

$$W_I = K_{I_0} \exp\left(\frac{\alpha_I q}{RT}\right) f_I ([A]) ([K]_0 - [AK]) \qquad\qquad (4)$$

$$W_{II} = K_{I_0} \exp\left(- \frac{1 - \alpha_{II} q}{RT}\right) f_{II} ([B]) [AK] \qquad\qquad (5)$$

Here K_{I_0} and K_{II_0}, coefficients independent of q and f_I ([A]) and f_{II} ([B]), are functions taking account of the dependence of stage rates on reactant concentration.

In the case of irreversibility of stage under stationary conditions, the overall reaction rate $W = W_I = W_{II}$. Then the intermediate product concentration is

$$[AK] = [K]_0 \bigg/ \left\{ 1 + G \exp\left(- \frac{1 + \alpha_I - \alpha_{II}}{RT} q \right) \right\} \qquad\qquad (6)$$

and the reaction rate is

$$W = \frac{K_{II_0} \exp\left(- \frac{1 - \alpha_{II}}{RT} q \right)}{1 + G \exp\left(- \frac{1 + \alpha_I - \alpha_{II}}{RT} q \right)} \qquad\qquad (7)$$

where

$$G = \frac{K_{II_0} \, f_{II} \, ([B])}{K_{I_0} \, f_I \, ([A])}$$

With increasing intermediate interaction heat (q), i. e. , the energy of the reactant A-catalyst bond, the reaction rate increases first and then decreases. The optimal value of the bond energy conforming to the reaction rate maximum is

$$q_{opt} = \frac{RT}{1 + \alpha_I - \alpha_{II}} \ln \frac{\alpha_I G}{1 - \alpha_{II}} \tag{8}$$

Substituting the optimal value of q from Eq. (8) into Eq. (6), we find the intermediate product concentration at the optimal energy of the reactant-catalyst bond

$$[AK] = \frac{\alpha_I [K]}{1 + \alpha_I - \alpha_{II}} \tag{9}$$

At $\alpha_I = \alpha_{II}$

$$q_{opt} = RT \ln \frac{\alpha}{1 - \alpha G}, \quad [AK]_{opt} = \alpha [K]_0$$

and at $\alpha_I = \alpha_{II} = 0.5$

$$q_{opt} = RT \ln G, \quad [AK]_{opt} = 0.5 [K]_0$$

The conclusion concerning the existence of an optimal energy of the reactant-catalyst bond is also valid for heterogeneous catalysis. In a general form, on the base of the BPS rule, this problem was treated by Temkin. Provided that the above simple two-phase scheme is realized on a solid catalyst surface, $[K]_0$ can be assumed equal to the number of identical sites on the catalyst surface taking part in intermediate interaction.

Then

$$[AK] / [K]_0 = \theta$$

can be treated as the active surface fraction covered by the chemisorbed reactant A, and according to Eq. (9) the surface fraction covered at the optimal bond energy is

$$\theta_{opt} = \frac{\alpha_I}{1 + \alpha_I - \alpha_{II}} \tag{10}$$

and at $\alpha_I = \alpha_{II}$

$$\theta_{opt} = \alpha$$

For catalytic oxidation reactions, the work of the Institute of Catalysis showed the governing role of the energy of the oxygen bond on the catalyst surface for its activity.[2]

For the determination of this energy various methods were
used; determination of the temperature dependence of the
catalyst oxygen equilibrium pressure at a given oxygen
content in an surface layer,[3]   direct calorimetric measurement while taking oxygen off the surface with a reducer, with
subsequent setting,[4]   and catalytic activity in relation to
isotopic exchange with molecular oxygen.[5]   Measurements of
oxygen bond energy on the surface of simple and complex
oxides gave adjusted results and showed that the catalytic
activity in relation to complete oxidation reactions increases
with decreasing oxygen bond energy on the catalyst surface.[6]
It points to the fact that the limiting stage of these reactions
is related to the oxygen breaking away from the catalyst.
In the case of selective oxidation, weakly bound oxygen is
unwanted as it leads to complete oxidation products; partial
oxidation results from the interaction with a more stably
bound oxygen on the surface.[7]   The concept of optimal
reactant bond energy is also used in catalytic hydrogenation,
where the hydrogen-catalyst bond energy can be determined
by electrochemical methods.[8]   In this direction a number of
investigations have been made for other catalytic reactions.
    Consider the problem of the influence of the energy of the
product-catalyst surface bond on a reaction rate. In contrast
with reactant chemisorption, the concept of optimal bond
energy of product chemisorption was not proposed. It was
beleived that as product chemisorption resulted in covering
a part of the catalyst surface, it is always harmful and the
less the heat of product chemisorption, i. e., the less the
product-catalyst bond energy, the better. It is not difficult
to show that this universally accepted notion is incorrect and
in any case does not possess any general significance. The
product adsorption heat influences not only the desorption
activation energy, and accordingly the rate of this stage, but
also the activation energy and rate of the chemical conversion stage. Therefore, just the same as for the heat of
reactant chemisorption, an optimal product chemisorption
heat exists and decreasing the chemisorption heat as
compared with this value results in decreasing the overall
catalytic reaction rate.
    Consider a simple reaction of heterogeneous catalysis

$$A + B \longrightarrow P$$

following the scheme which includes the reactant A chemisorption stage

$$A + \square \longrightarrow A_{ad} \tag{I}$$

the interaction of A_{ad} with the second reactant B results in the formation of an adsorbed product

$$A_{ad} + B \longrightarrow P_{ad} \tag{II}$$

and product desorption

$$P_{ad} \longrightarrow P + \square \tag{III}$$

Here $\square$ means a region free for chemisorption on the catalyst active surface.

The energetic pathway of the reaction includes three maxima corresponding to the three reaction stages and two minima corresponding to the chemisorbed reactant A and product (Fig. 1).

If the reactant chemisorption heat is q_1 and the product chemisorption heat is q_2, then the energies of the above stage activation in accordance with BPS rule will be

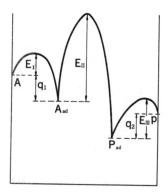

Fig. 1. Reaction pathway of intermediate interaction with catalyst. E_I, E_{II}, E_{III} are energies of activation of stages I, II, III, q_1 and q_2 are heats of reactant and product chemisorption.

$$E_1 = E_{I_0} - \alpha_1 q_1$$

$$E_{II} = E_{II_0} + (1 - \alpha_2) q_1 - \alpha_3 q_2 \qquad (11)$$

$$E_{III} = E_{III_0} + (1 - \alpha_4) q_2$$

The first stage activation energy decreases with increasing reactant chemisorption heat, the second stage activation energy increases with increasing reactant chemisorption heat and decreases with increasing product chemisorption heat, and the third stage activation energy increases with increasing product chemisorption heat. From this it is already clear that the influenece of the reactant and product heats of chemisorption act symmetrically, and optimal values for both q_1 and q_2 should exist.

For simplicity in the following calculations all coefficients α_i are considered to be equal. The reaction under consideration and its stages are assumed to be irreversible and all active sites on the catalyst surface to be identical.

The reactant chemisorption stage rate is

$$W_I = K_{I_0} \exp\left(-\frac{E_{I_0} - \alpha q_1}{RT}\right) f \left[A\right]) (1 + \theta_1 - \theta_2)$$

$$= K_I \exp\left(\frac{\alpha q_1}{RT}\right) (1 - \theta_1 - \theta_2)$$

The chemical conversion stage rate is

$$W_{II} = K_{II_0} \exp\left\{\frac{-E_{II_0} + (1-\alpha) q_1 - \alpha q_2}{RT}\right\} f \left(\left[B\right]\right) \theta_1$$

$$= K_{II} \exp\left\{\frac{\alpha q_2 - (1-\alpha) q_1}{RT}\right\} \theta_1$$

The product desorption stage rate is

$$W_{III} = K_{III_0} \exp\left\{\frac{-E_{III_0} + (1-\alpha) q_2}{RT}\right\} \theta_2$$

$$= K_{III} \exp\left\{-\frac{(1-\alpha) q_2}{RT}\right\} \theta_2$$

Here θ_1 represents the catalyst surface fraction covered by chemisorbed reactant, θ_2 is the surface fraction covered with chemisorbed product, $f([A])$ and $f([B])$ are the functions reflecting dependence of the first and second stage rates on the reactant concentrations, and K_I, K_{II}, and K_{III} are coefficients independent of the reactant and product chemisorption heats.

The overall reaction rate under stationary state conditions is

$$W = W_I = W_{II} = W_{III}$$

From this it follows

$$K \exp\left(\frac{\alpha q_1}{RT}\right)(1-\theta_1-\theta_2) = K_{II}\exp\left\{\frac{\alpha q_2-(1-\alpha)q_1}{RT}\right\}\theta_1$$

$$= K_{III}\exp\left\{\frac{-(1-\alpha)q_2}{RT}\right\}\theta_2 \tag{12}$$

$$\theta_1 = 1/\left\{1+\frac{K_{II}}{K_I}\exp\left(\frac{\alpha q_2-q_1}{RT}\right)+\frac{K_{II}}{K_{III}}\exp\left(\frac{q_2-(1-\alpha)q_1}{RT}\right)\right\} \tag{13}$$

$$\theta_2 = \theta_1 \frac{K_{II}}{K_{III}}\exp\left\{\frac{q_2-(1-\alpha)q_1}{RT}\right\}$$

and the reaction rate is

$$W = \frac{K_{II}\exp\left(\dfrac{\alpha q_2-(1-\alpha)q_1}{RT}\right)}{1+\dfrac{K_{II}}{K_I}\exp\left(\dfrac{\alpha q_2-q_1}{RT}\right)+\dfrac{K_{II}}{K_{III}}\exp\left(\dfrac{q_2-(1-\alpha)q_1}{RT}\right)}$$

The optimal values of q_1 and q_2 can be determined from the conditions:

$$\left(\frac{\partial W}{\partial q_1}\right)_{q_1} = \text{const} = 0, \quad \left(\frac{\partial W}{\partial q_2}\right)^2_{q_2} = \text{const} = 0$$

The first condition results in the relation

$$\frac{\alpha}{1-\alpha}\cdot\frac{K_{II}}{K_I}\exp\frac{\alpha q_2}{RT} = \exp\frac{q_{1opt}}{RT}$$

or

$$q_{1\,opt} = RT \ln \left\{ \frac{K_{II}}{K_I} \cdot \frac{\alpha}{(1-\alpha)} \right\} + \alpha q_2 \qquad (14)$$

The optimal degree of covering of the catalyst surface by a chemisorbed reactant is

$$\theta_{1\,opt} = 1 / \left\{ 1 + (1-\alpha)/\alpha + \theta_2 / \theta_{1\,opt} \right\} = \alpha\,(1-\theta_2)$$

The optimal covering by a reactant corresponds to the fraction of part α from the catalyst surface which is free of a chemisorbed product.

From the second condition we find

$$\frac{\alpha}{1-\alpha} \cdot \frac{K_{III}}{K_{II}} \exp \frac{(1-\alpha)q_1}{RT} = \exp \frac{q_{2\,opt}}{RT}$$

or

$$q_{2\,opt} = RT \ln \left(\frac{K_{III}}{K_{II}} \cdot \frac{\alpha}{1-\alpha} \right) + (1-\alpha)q_1$$

In combination with Eq. (14) for $q_{1\,opt}$ we find

$$q_{1\,opt} = \frac{RT}{1-\alpha(1-\alpha)} \left\{ \alpha \ln K_{III} + (1-\alpha)\ln K_{II} - \ln K_I + \frac{(1+\alpha)\ln \alpha}{1-\alpha} \right\}$$

and

$$q_{2\,opt} = \frac{RT}{1-\alpha(1-\alpha)} \left\{ \ln K_{III} - \alpha \ln K_{II} - (1-\alpha)\ln K_I + \frac{(2-\alpha)\ln \alpha}{1-\alpha} \right\}$$

The optimal surface coverage by a chemisorbed reaction product is

$$\theta_{2\,opt} = \theta_{opt}\ \alpha / (1-\alpha)$$

As

$$\theta_{1\,opt} = \alpha\,(1-\theta_{2\,opt})$$

$$\theta_{1\,opt} = \frac{\alpha(1-\alpha)}{1-\alpha+\alpha^2} \qquad \text{and} \qquad \theta_{2\,opt} = \frac{\alpha^2}{1+\alpha+\alpha^2}$$

In the case where $\alpha = 0.5$

$$\theta_{1\,opt} = \theta_{2\,opt} = \frac{1}{3}$$

Thus, the maximal catalytic reaction rate is reached under stationary state conditions at a particular and quite appreciable catalyst surface coverage by chemisorbed product. Decreasing the product chemisorption heat results not only in a decrease in the surface coverage by product but also in a decrease in the catalytic reaction rate. In the specific case of $\alpha = 0.5$ at the optimal energies of the reactant-product-catalyst bonds the catalyst surface is equally distributed between free sites, sites covered by chemisorbed reactant (θ_1), and sites covered by chemisorbed product (θ_2).

The relations found are also applicable for more complex catalytic reactions of the type

$$\Sigma A_i + \Sigma B_i \longrightarrow \Sigma P_i + \Sigma \theta_i + \Sigma R_i$$

including a chemisorption stage

$$\Sigma A_i + \square \longrightarrow X + \Sigma P_i$$

a chemical conversion stage

$$X + \Sigma B_i \longrightarrow Y + \Sigma \theta_i$$

and a product desorption stage

$$Y \longrightarrow \Sigma R_i + \square$$

Here X represents the product of surface intermediate interaction with catalyst of reactants A_i, and X is chemisorbed by catalyst with products of reaction R_i.

The above conclusion about the possibility of favourable influence of the product chemisorption energy of a catalytic reaction on its rate is also valid in the case where reversibility of the reaction and its separate stages holds.

As an example, we will consider the same reaction assuming the reversibility of all stages. We will assume that the processes of reactant chemisorption and desorption and product desorption and adsorption are much quicker than chemical conversions, and in fact that adsorption equilibrium is reached. Then, assuming the identity of all catalytically-active surface sites we find

$$\theta_1 = \frac{B_{1_0}[A]\exp[q_2/RT]}{1+B_{1_0}[A]\exp(q_1/RT)+B_{2_0}[P]\exp(q_2/RT)}$$

$$\theta_2 = \frac{B_{2_0}[P]\exp(q_2/RT)}{1+B_{1_0}[A]\exp(q_1/RT)+B_{2_0}[P]\exp(q_2/RT)}$$

and the reaction rate is

$$W = \frac{K_{II}\exp\left\{\dfrac{\alpha_2 q_2-(1-\alpha_1)q_1}{RT}\right\}B_{1_0}[A]\exp\left(\dfrac{q_1}{RT}\right)}{1+B_{1_0}[A]\exp\left(\dfrac{q_1}{RT}\right)+B_{2_0}[P]\exp\left(\dfrac{q_2}{RT}\right)}$$

The conditions for maximum q_1 give

$$q_{opt} = RT\ln\left(\frac{\alpha_2}{1-\alpha_2}\right)\frac{1-B_{1_0}[A]\exp(q_1/RT)}{B_{2_0}[P]}$$

and for the optimum of the product bond energy

$$q_{opt} = RT\ln\left(\frac{\alpha_2}{1-\alpha_1}\right)\frac{1-B_{1_0}[A]\exp(q_1/RT)}{B_{2_0}[P]}.$$

Thus in this case at a given q_1 the reaction rate increases with increasing q_2, reaching a maximum at $q_1 = q_2$.

The integral optimal values of q_1 and q_2 corresponding to the absolute maximum of the reaction rate are possible only under the condition $\alpha_1 + \alpha_2 < 1$.

In this case

$$q_{1opt} = \frac{RT\ln\alpha_1}{B_{1_0}[A](1-\alpha_1-\alpha_2)}$$

and

$$q_{2opt} = \frac{RT\ln\alpha_2}{B_{2_0}[P](1-\alpha_1-\alpha_2)}$$

The degrees of coverage at optimal bond energies are:

$$\theta_{1\,opt} = \frac{\alpha_1/(1-\alpha_1-\alpha_2)}{1+1/(1-\alpha_1-\alpha_2)+\alpha_2/(1-\alpha_1-\alpha_2)} = \alpha_1$$

and

$$\theta_{2\,opt} = \alpha_2$$

The energy of the product-catalyst bond will not influence the energy of the chemical conversion stage activation except in that highly improbable case when the chemical conversion directly results in all products separating into a gas phase, and only then does their adsorption take place.

In conclusion I would like to emphasize that the energy of the reaction product-catalyst bond influences the reaction rate quite as much as the energy of reactant chemisorption, and therefore should be considered as an equally important parameter in determining catalytic activity.

References

1) M.I. Temkin, Zh. Fiz. Khimii, 31, 3 (1957); M.I. Temkin and S.L. Kiperman, Zh. Fiz. Khimii, 21, 927 (1947).

2) G.K. Boreskov, Kinetika i Kataliz, 8, 1020 (1967).

3) G.K. Boreskov, V.A. Sazonov, and V.V. Popovsky, Kinetika i Kataliz, 9, 307 (1968).

4) G.K. Boreskov, Yu.D. Pankratjev, V.I. Solovjev, V.V. Popovsky, and V.A. Sazanov, DAN CCCP 184, 611 (1969).

5) G.K. Boreskov, A.P. Dzisyak, and L.A. Kasatkina, Kinetika i Kataliz, 4, 388 (1963), G.K. Boreskov, Advan. Catalysis, 15, 285 (1964).

6) G.K. Boreskov, V.V. Popovsky, and V.A. Sazonov, in Trudy IV Mezhdunarodnogo Kongressa po Katalizu 1968 goda, tom. 1, 343, Izd. Nauka, Moskva, 1970.

7) B.I. Popov, B.N. Bibin, and G.K. Boreskov, Kinetika i Kataliz, 9, 796 (1968).

8) D.V. Sokoljsky, Optimaljnye Katalizatory gidlirovaniya v rastbopakh, Izd. Nauka Kaz. CCP, Alma-Ata (1970).

On the Activities of Oxide Catalysts for the Allylic Oxidation of Olefins

T. Seiyama, M. Egashira and M. Iwamoto

Kyushu University, Fukuoka

INTRODUCTION

Catalytic oxidation of olefins is one of the most interesting subjects in heterogenous catalysis. Many investigations have been done with respect to the origins of the activities and selectivities of various binary oxide catalysts, such as the Bi_2O_3-MoO_3 system, in the oxidation of propylene to acrolein or acrylonitrile, and of n-butenes to butadiene. However, there are many unsolved problems, including the oxidation of olefin to aromatics (or diolefins) over bismuth phosphate catalysts[1]. For example, propylene is oxidized to benzene, and iso-butene to p-xylene or 2, 5-dimethyl-1, 5-hexadiene. This aromatization reaction was also reported later over the Bi_2O_3-SnO_2 system by Ohdan et al.[2] and over Tl_2O_3 or In_2O_3 by Trimm et al.[3]

The reaction was found to proceed through several consecutive steps in the presence of oxygen:[1] the oxidative dehydrogenation of olefin to an allylic intermediate, dimerization of the allylic intermediates to diolefins such as 1, 5-hexadiene, and aromatization of the diolefins. The participation of allylic species in the reaction was also confirmed in the oxidation of n-butenes by the facts that butadiene, which is known to be formed via the allylic intermediate, was obtained together with small amounts of aromatics and diolefins, and that the distribution of products was the same for the oxidation of three kinds of n-butenes.

Thus the reaction can appropriately be designated as oxidative dehydroaromatization and classified as an allylic type oxidation of olefins[4], as is acrolein formation. The reaction is clearly distinguished from the well-known dehydrocyclization of C_6 and higher paraffins (or olefins), and from the dehydrocyclodimerization of C_3-C_5 paraffins (or olefins), which was found by Csicsery[5].

Since oxidative dehydroaromatization occurs competitively with acrolein formation, a systematic investigation of the

catalytic properties of various oxides for each allylic oxidation is expected to provide useful information about the
origins of activity and selectivity. In the present work, the
allylic oxidation of propylene to benzene or acrolein was
examined, using twenty-four metal oxides as catalysts. The
intention was to seek regularities in the catalytic activities
of oxides, and to determine the catalyst properties responsible for selecting whether propylene will be converted to
benzene by the coupling of two allylic intermediates, or to
acrolein by the addition of an oxygen atom. Studies were
also extended to catalysts of binary oxide systems containing
bismuth or tin, to investigate the promoting effect of the
second component oxide.

EXPERIMENTAL

Catalysts

Twenty-four metal oxides and two series of binary oxide
systems containing bismuth or tin oxide were used as catalysts. They were prepared as described below and calcined
at 600°C for 5 hrs before used. The structure of the catalyst was identified by X-ray diffraction and surface area was
measured by the BET method.

The metal oxides, except V_2O_5 and SnO_2, were commercial materials of extra-pure grade. V_2O_5 was obtained by
calcination of ammonium metavanadate at 200°C for 9 hrs
and again for 1 hr after addition of a small amount of nitric
acid; it was converted into a crystalline state by calcining at
800°C for 18 hrs. SnO_2 was prepared by calcination of
stannic acid, which was precipitated from stannic chloride
solution by ammonia, at 600°C for 5 hrs.

A series of bismuth oxide-metal oxide systems were
prepared and are listed in Table I. They were obtained by
evaporation to dryness from a mixed solution or suspension
of bismuth nitrate and the ammonium salt of the respective
oxy acid. However, basic bismuth sulfate was obtained by
the thermal decomposition of $Bi_2(SO_4)_3$, and bismuth titanate
by calcination of a solid mixture of Bi_2O_3 and TiO_3.

Stannic oxide-metal oxide systems were prepared by
impregnation of pure SnO_2 with alkali or alkaline earth
hydroxide solution, or with ammonium salt solution of oxy
acid, followed by evaporation.

Table I.   Bismuth Salt Catalysts.

Oxy Salt	Bi/M	Composition of Catalyst (by X-ray diffraction)
Phosphate	2/1	$3Bi_2O_3 \cdot P_2O_5$, $2Bi_2O_3 \cdot P_2O_5$, three forms of $BiPO_4$ (hexagonal, monazite, high temperature), and small amounts of Bi_2O_3
	1/1[a]	high temperature form $BiPO_4$
	1/1[b]	monazite form $BiPO_4$
Arsenate	1/1	monazite form $BiAsO_4$
Basic sulfate	2/1	$(BiO)_2SO_4$
Titanate	1/1	$Bi_4 (TiO_4)_3$, and small amounts of Bi_2O_3 and TiO_2
Molybdate	2/1	$(BiO)_2MoO_4$ (koechlinite)
	1/1	$Bi(BiO) (MoO_4)_2$
	2/3	$Bi_2 (MoO_4)_3$

a   Calcined at 600°C for 5 hrs.
b   Calcined at 500°C for 5 hrs.

Apparatus and Procedures

Oxidation was carried out in a flow system at atomospheric pressure, using a 4-8 mm ID Pyrex or quarz glass tube reactor with a fixed bed. A typical feed gas composition was 9% propylene, 18% oxygen, and 73% nitrogen. The feed gas was passed over catalyst granules of 48-80 mesh under conditions of contact time $W/F = 10^{-3} \sim 50$ g. sec /ml and temperature of $300 \sim 600°C$. The temperature was measured by a thermocouple placed in the catalyst bed. Under these conditions, a steady-state was attained within 1 hr, and this was held for 5 or 10 hrs. Data were usually obtained 2 or 3 hrs after the reaction was started. Feed and exit gases were analyzed by gas chromatography, using Porapak Q column for propylene and carbon dioxide, Molecular Sieve

5A for oxygen and carbon monoxide, and 10% PEG 1000 on
Chromosorb for acrolein, benzene and 1, 5-hexadiene.

RESULTS

Catalytic Properties of Metal Oxides
    In the examination of the catalytic properties of metal
oxides for allylic oxidations, regard must be paid to the
results of preliminary experiments. Acrolein is formed
above 350°C and becomes significant at 450~500°C, but
production decreases above 500°C due to consecutive oxidation to carbon dioxide. Meanwhile, benzene formation occurs
above 450°C and markedly increases at 500~550°C. From
the dependence of yield on contact time, acrolein or 1, 5-
hexadiene is indicated to be a primary product of consecutive
reaction, whereas benzene may be a secondary product.
    In the first place, in order to find active oxides for oxida-
tive dehydroaromatization, the catalytic behavior of various
oxide catalysts was examined at a contact time $W/F = 2.0$ g.
sec/ml and at a temperature of 500°C. The results show
that several oxides such as ZnO, In_2O_3, SnO_2, and CdO are
active as summarized in Table II.
    These results, however, are not adequate for a discussion
on regularities in catalytic behavior, as the conversion levels
were quite different. For this purpose, data at constant total
conversion were obtained from the dependence of reactivity
on contact time. It is kinetically desirable that the conversion level be lower than 10%. At such a level, however, it
is expected that little benzene formation would be observed,

Table II.   Active Oxides for Oxidative Dehydroaromatization.
(C_3H_6 = 9%, O_2 = 18%, T = 500°C, W/F = 2.0 g.sec/ml)

Oxides	Conversion of propylene (%)	Selectivity (%)		
		for $CO_2 + CO$	for C_3H_4O	for C_6H_6
ZnO	66.1	86.7	tr.	12.0
In_2O_3	57.7	86.4	2.1	10.9
SnO_2	63.4	81.5	1.0	17.5
CdO	62.3	88.1	0	11.7

because benzene is a secondary product from propylene. In
fact, under such condition, the amount of benzene formed
was very small over SnO_2 and CdO. For this reason, a
level of 25% conversion was adopted in the present study,
and the rate obtained was taken to roughly indicate the
relative order of catalytic activities of various oxides, as
the conversion showed a linear relation against contact time
up to the 25% level in many oxides. The results are shown
in Table III. From Table III, the catalytic properties of
metal oxides are appropriately classified into the following
four groups:

(A) Acrolein formers; $MoO_3 > Sb_2O_4 > V_2O_5 > TiO_2 \approx Fe_2O_3$
$> SiO_2 > WO_3$, Al_2O_3,

(B) Benzene or 1,5-hexadiene formers; $ZnO > Bi_2O_3$
$> In_2O_3 > Sn\ O_2 > CdO$,

(C) Highly active CO_2 formers; $CuO \approx Co_3O_4 > NiO$
$> Cr_2O_3 > MnO_2 \approx GeO_2$,

(D) Low active CO_2 formers; CaO, MgO, CeO_2, (PbO),
(Tl_2O_3).

Catalysts of Bismuth Salts of Oxy Acids

To obtain information on the origins of activity of bismuth
phosphate, the oxidation of propylene was studied using
various metal phosphates——Ca^{2+}, Cr^{3+}, Fe^{3+}, Co^{2+}, Ni^{2+},
Cu^{2+}, Zn^{2+} and Ce^{3+}——and various bismuth salts listed in
Table I.

Among the phosphates examined, only three bismuth
phosphates showed activity for benzene formation. On the
other hand, some of the bismuth salt catalysts were active
as shown in Table IV. These facts indicate that the presence of phosphate ion is not a sufficient requirement for the
oxidative dehydromatization. As seen from Table IV, most
of the bismuth salts of oxy acids are active for the allylic
oxidation of propylene. Their activities are far higher than
expected from the results in Table III. The activity of
bismuth salts is seen to arise from the structural specificity
as an oxy-salt. In these systems, it can reasonably be
considered that the Bi^{3+} ion is the active component for
allylic oxidation, and that its activity is promoted by adjacent oxy anions, or M^{n+}——O^{2-} groups.

In regard to the selectivity, the influence of oxy anions or
M^{n+}——O^{2-} groups must be significant, as well as the role of
Bi^{3+}, because the selectivity can be changed from primarily

benzene to primarily acrolein according to the species of oxy anion.

Table III.   Catalytic Properties of Metal Oxides for the Oxidation of Propylene at the Constant Conversion Level of 25%.

$(C_3H_6 = 9\%, \ O_2 = 18\%, \ T = 500°C)$

Oxides		Surface area (m^2/g)	Contact time[a] $(m^2 \cdot sec/ml)$	Conversion of propylene (%)	Selectivity[b] CO₂ + CO (%)	C₃H₄O (%)	C₆H₆ (%)
A	MoO₃ (n)	1.55	30.8	25.0	74.6	25.1	0
	Sb₂O₄ (n)	1.66	32.0	25.0	63.3	16.3	0
	V₂O₅ (n)	1.37	0.121	25.0	81.8	14.2	0
	TiO₂ (n)	1.67	20.4	25.0	85.8	12.0	0.3
	Fe₂O₃ (n)	3.05	0.174	25.0	86.9	12.2	tr..
	SiO₂	41.8	1.25	25.0	46.7	8.6	tr.
	WO₃ (n)	4.25	12.3	25.0	90.0	6.3	0.9
	Al₂O₃ (i)	--		4.3[c]	38.5[c]	42.1[c]	0[c]
B	ZnO (n)	7.28	0.167	25.0	49.4	0.5	48.2
	Bi₂O₃	0.105	1.19	25.0	89.4	0	9.6[d]
	In₂O₃ (n)	0.74	0.0029	25.0	93.4	0.8	4.1
	SnO₂ (n)	15.3	0.073	25.0	91.5	3.2	2.5[e]
	CdO (n)	2.13	1.12	25.0	98.5	0.8	0.1
C	CuO[f] (n)	0.85	0.0021	25.0	92.5	3.8	0
	Co₃O₄ (p)	0.87	0.0012	25.0	96.2	2.2	0
	NiO (p)	0.71	0.0085	25.0	75.7	2.3	0
	Cr₂O (p)	2.80	0.364	25.0	69.3	2.1	4.0[d]
	MnO₂ (p)	37.3	1.79	25.0	99.7	0.3	0
	GeO₂ (n)	0.83	1.07	25.0	99.9	tr.	0
D	CaO (n)	27.5	12.9	25.0	86.5	0	0
	MgO (i)	--		9.0[c]	94.0[c]	tr.[c]	0[c]
	CeO₂	--		8.3[c]	94.0[c]	tr.[c]	0[c]
	PbO[f] (p)	--		4.1[c]	69.5[c]	tr.[c]	0[c]
	Tl₂O₃[f] (n)	--		18.7[c]	96.7[c]	tr.[c]	0[c]

a   Conversion levels reached 25% at these contact times.
b   Other products were ethylene, acetaldehyde, and so on.
c   Data at the constant contact time $\underline{W/F}$ = 2.0 g. sec/ml.
d   Values for 1,5-hexadiene.
e   0.7 for benzene, and 1.8 for 1,5-hexadiene.
f   These oxides were partially reduced under the working conditions.

Batist et al.[6] and Peacock et al.[7] assumed that the adsorption sites for the allylic intermediate over bismuth molybdate catalysts would be Mo^{6+} ions. Over this catalyst, however, it is more appropriate to suppose that Bi^{3+} ions provide the adsorption site under the influence of the adjacent oxy anion, MoO_4^{2-}, as in the case of other oxy-salts.

Table IV. Catalytic Properties for the Allylic Oxidation of Propylene, and Acid Strength of Bismuth Salt Catalysts.

(C_3H_6 = 9%, O_2 = 18%, T = 500°C, W/F = 2.0 g. sec/ml)

| Catalysts | Conversion of propylene (%) | Selectivity (%)[a] | | Acid strength[b] H_0 |
		for C_6H_6	for C_3H_4O	
$2Bi_2O \cdot P_2O_5$ (Bi/P = 2)	80.4	49.0	0	+7.1~+6.8
$BiAsO_4$ (Monazite)	69.0	33.8	5.8	+6.8~+4.0
$BiPO_4$ (High temp.)	79.8	26.9	6.6	+6.8~+4.0
$Bi_2O_3 \cdot 2TiO_2$ (Bi/Ti = 1)	45.7	18.0	0.3	+7.1~+6.8
$(BiO)_2SO_4$	61.0	10.1	4.0	+6.8~+4.0
$BiPO_4$ (Monazite)	43.0	9.1	38.6	+1.5~-3.0
$(BiO)_2MoO_4$ (Bi/Mo = 2)	37.7	0	66.1	+3.3~+1.5
$Bi(BiO)(MoO_4)_2$ (Bi/Mo = 1)	23.7	0	91.7	+1.5~-3.0
$Bi_2(MoO_4)_3$ (Bi/Mo = 2/3)	5.9	0	94.9	+1.5~-3.0

a Other product was mainly carbon dioxide.
b Hammett's indicators used were bromothymol blue (pK_{BH} = +7.1), neutral red (+6.8), phenylazonaphthylamine (+4.0), butter yellow (+3.3), benzeneazodiphenylamine (+1.5), and dicinnamalacetone (-3.0).

Stannic Oxide Promoted by Basic or Acidic Oxides

The activity of stannic oxide for the allylic oxidation of propylene was found to be promoted by basic or acidic oxides. Figure 1 shows the promoting effects of basic Na_2O and acidic P_2O_5 on benzene formation; P_2O_5 caused an

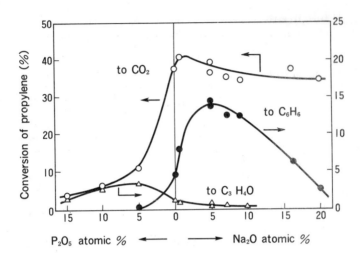

Fig. 1.   The promoting effects of Na_2O and P_2O_5 on the catalytic behavior of stannic oxide.
(C_3H_6 = 12%, O_2 = 24%, T = 550°C, W/F = 2.0 g. sec/ml)

increase in acrolein formation despite the decrease of total activity.   Other alkali oxides and acidic oxides such as MoO_3 and WO_3 were also found to have similar promoting effects.

DISCUSSION

Catalytic Activities of Metal Oxides

The order of activity of catalysts is often discussed by referring to thermodynamic parameters such as the heat of formation of catalyst oxide.   In the oxidation of propylene, Moro-oka et al.[8] obtained a good correlation between the catalytic activities and the heats of formation of the oxides, but they excluded most of the oxides belonging to groups A and B, which are of interest here.

Figure 2 shows the correlation, where the rates of complete oxidation (CO_2 + CO) obtained from Table 3 are plotted against the heats of formation ($-\Delta H_f^0$) of oxides per

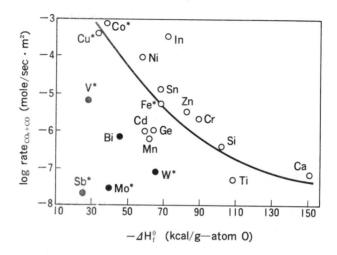

Fig. 2. Correlation between the rate of complete oxidation (CO_2 + CO) and the heat of formation ($-\Delta H_f^0$) of oxide per g-atom of oxygen.

g-atom of oxygen. * The rates obtained here may be qualitatively accepted as a parameter of catalytic activity, though agreement is not always complete. As seen in the figure, a fairly good correlation is observed among the oxides plotted by open circles (1st group), while the oxides plotted by black circles (2nd group) deviate markedly from the correlation line. In the activity patterns for allylic oxidations, a similar correlation is also seen, though it is a little poorer, as shown in Fig. 3 where the total rates of two types of allylic oxidation are used as the activities.

The trend in Figs. 2 and 3 is consistent with that of Gelbshtein et al.[9] They divided oxides into two groups, from the relation between the activation energy (E) of complete oxidation of 1-butene over the oxides and the activation energy (E_0) of isotopic exchange of the oxides with molecular oxygen. According to them, the 1st group oxides in Figs. 2

* For the oxides marked with asterisk, the heats of reaction of the oxidation-reduction cycle, $MO_{x-1} + \frac{1}{2}O_2 \rightleftharpoons MO_x$, were used instead of $-\Delta H_f^0$.

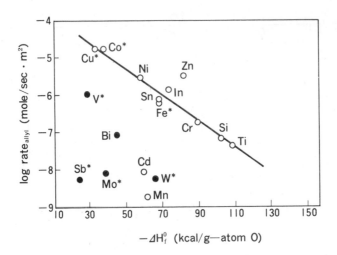

Fig. 3.   Correlation between the total rate of allylic
oxidations and the heat of formation ($-\Delta H_f^0$) of oxide per
g-atom of oxygen.

and 3 have E values equal to E_0, while the low activity 2nd
group oxides have E values much smaller than E_0.   The
activation energy E characterizes the metal-oxygen bond
strength on the surface, and in general corresponds to the
parameter $-\Delta H_f^{0}$,[10] although for the 2nd group oxides this
correlation between E_0 and $-\Delta H_f^0$ is absent. For the 1st
group, therefore, it may be concluded that the metal-oxygen
bond strength on the surface is the factor determining the
catalytic activity for both the complete and allylic oxidations,
and that the rate is controlled by the rupture of this bond.
　　On the other hand, the situation for the 2nd group is
characterized by the following points: (1) low catalytic
activities for both oxidations, when compared with the 1st
group, (2) larger E_0 values than expected from the correlation between E and $-\Delta H_f^0$ in the 1st group, indicating that
the surface oxygen bonds are stronger than those in the oxide
bulk, (3) the lack of linear dependence of E on E_0, and (4)
the absence of the correlation between activity and $-\Delta H_f^0$.
It may be supposed that low catalytic activities (1) are
caused by the strong surface oxygen bond (2). However, as

the rupture of the metal-oxygen bond does not control the rate, as suggested from (3) and (4), this supposition is ruled out. The low activity of the 2nd group is well accounted for on the assumption that hydrogen abstraction from the allylic position of the olefin would play a predominant role for controlling the rate, as in the case of Bi_2O_3-MoO_3 catalysts. However, details about this are not clear yet.

Selectivity of Catalyst

The selectivity of oxide catalysts is a matter of interest. First we will consider the total selectivity for allylic oxidations. Comparison of the two correlation curves in Figs. 2 and 3 shows that the decrease in catalytic activity with $-\Delta H_f^0$ for partial oxidation is less than that for complete oxidation. This leads to an expectation that the oxides with larger $-\Delta H_f^0$ exhibit higher selectivities than those with smaller $-\Delta H_f^0$. It means that the strength of the M-O bond is one of the factors affecting the selectivity of catalyst. Though the oxides of the 1st and 2nd groups behave in different ways in the oxygen isotopic exchange reaction, the activation energy E_0 of the exchange can be used as a common parameter for both groups. Figure 4 shows the selectivities plotted against the E_0 values.[9-11] The distinction between the 1st and 2nd groups disappears, and the selectivities are observed to be correlated with the E_0 values thoughout both groups of oxides, except for the extraordinarily high selectivity of ZnO. These results may be interpreted as showing that the M-O bond strength participates in the selection of the consecutive oxidation paths from the adsorbed allylic intermediate.

It will be also useful to discuss the results on the basis of the semiconductor theory of catalysis, which was developed by Vol'kenshtein. It is noticeable in the classification in Table III that most of oxides in groups A and B are n-type semiconductors, while those in group C are chiefly of p-type. A similar trend is seen in semiconductor-catalyzed reactions like N_2O decomposition.[12] In general, molecules which act as electron-acceptors on the semiconductor surface, for example oxygen, adsorb to larger extents on p-type semiconductors than on n-types, while electron-donating molecules such as propylene show opposite tendencies. Such a situation would result in a difference in the concentration ratio of (propylene)/(oxygen) over the oxide surface according to its semiconductive nature. Thus, the ratio is larger over

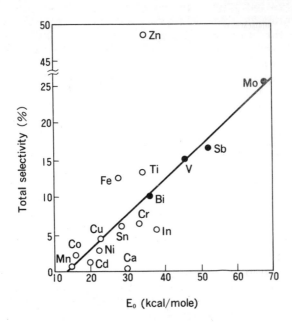

Fig. 4.   Correlation between the total selectivity for allylic oxidations and the activation energy (E_0) of oxygen isotopic exchange.

n-type oxides than over p-types, which would make it less preferable for propylene to be oxidized completely over the former oxides.

Additionally, it is notable that acrolein producers coincide with the oxides containing metal ions of electronic configuration d^0 or d^5, whereas benzene producers coincide with those of d^{10}. This may be explained with respect to the bond strength between the allylic intermediate and the metal ion.

Course of Allylic Oxidation to Acrolein or Benzene

It is also interesting whether the allylic oxidation would proceed to acrolein formation or to benzene formation. No correlation was found with parameters concerning the 'reactivity' of the surface oxygen anion, such as the heat of formation of the oxide and the activation energy of oxygen isotopic exchange. However, the acid-base properties of the oxide, such as the electronegativity and the ionic potential of the metal ion, seem to have a significant effect. Figure 5 shows the dependence of the selectivity on the electronegativity of the metal ion (Xi). For acrolein formation, high selectivity is seen on the side of larger value of Xi (acidic oxides), while benzene or 1, 5-hexadiene formation is favorable over oxides of rather smaller value of Xi (6.5 < Xi < 10) (slightly acidic or slightly basic).

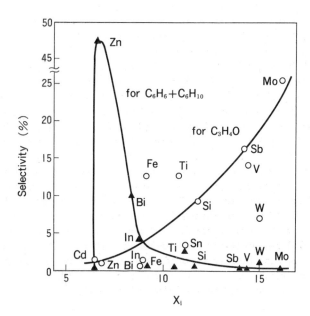

Fig. 5. The dependence of the selectivity for acrolein or for benzene + 1, 5-hexadiene (data in Table III) on the electronegativity (Xi) of the metal ion.

The correlation is supported by results over binary oxide catalysts. In the case of bismuth salt catalysts, the acid-base properties were measured using Hammett's indicators. Results are also shown in Table IV. Less acidic catalysts such as that with $Bi/P = 2$ are benzene-forming, while more acidic ones such as bismuth molybdate are acrolein-forming. In the case of stannic oxide catalysts, moreover, benzene formation is highly enhanced by the addition of basic oxides, while acrolein formation is promoted by acidic oxides.

The promoting effects of oxy anions on the Bi^{3+} ion, and of basic or acidic oxides on SnO_2, probably include changes in the electronic state or the acid-base properties of adsorption sites Bi^{3+} (or Sn^{4+}) through the bond Bi^{3+} (or Sn^{4+})——O^{2-}—M^{n+}. According to the principle of electronegativity equalization, if M^{n+} is more electronegative than Bi^{3+} (or Sn^{4+}), and if the ratio of $[M^{n+}]/[Bi^{3+}$(or $Sn^{4+})]$ is larger, the adsorption site becomes more acidic, and vice versa.

The acidity of the metal ion can be considered to influence the course of the reaction in two ways, because it acts as an adsorption site for an allylic intermediate or an oxygen atom. It may alter the electronic properties of adsorbed allylic species, or it may affect the metal-oxygen bond strength. However, selectivity for benzene or acrolein formation may be explained qualitatively in terms of the electronic state of the allylic intermediates, as shown below. An allylic intermediate has been indicated to be neutral or slightly positive.[13] On acidic adsorption sites, because of the large electronegativity of the metal ion, an electron of the allylic species would be localized near a metal ion. Thus the intermediate would become cationic and so susceptible to nucleophilic addition of an oxygen anion. The more acidic the catalysts, therefore, the higher becomes its selectivity for acrolein formation. On the other hand, on less acidic sites the allylic species would retain its radical-like character. It is reasonable to accept that benzene would be formed from the radical-like species, since low electrostatic repulsion would facilitate the dimerization step of two allylic intermediates.

References

1) T. Sakamoto, M. Egashira, and T. Seiyama, J. Catalysis, 16, 407 (1970).
2) K. Ohdan, T. Ogawa, S. Umemura, and K. Yamada, Kogyo Kagaku Zasshi (J. Chem. Soc. Japan), 73, 842 (1970).
3) D.L. Trimm and L.A. Doerr, Chem. Commun., 1970 1303.

4) H.H. Voge and C.R. Adams, Advan. Catalysis, 17, 151 (1967).

5) S.M. Csicsery, J. Catalysis, 17, 207, 216, 315, 323 (1970).

6) Ph.A. Batist, B.C. Lippens, and G.C.A. Schuit, J. Catalysis, 5, 55 (1966).

7) J.M. Peacock, M.J. Sharp, A.J. Parker, P.G. Ashmore, and J.A. Hochey, J. Catalysis, 15, 379 (1969).

8) Y. Moro-oka and A. Ozaki, J. Catalysis, 5, 116 (1966).

9) A.I. Gelbshtein, S.S. Stroeva, Yu.M. Kakshi, and Yu.A. Mischenko, in 4th Int. Cong. Catalysis, Moscow, 1968, Preprint of paper No.22.

10) G.K. Boreskov, Advan. Catalysis, 15, 285 (1964).

11) E.R.S. Winter, J. Chem. Soc., A-1968, 2889; A-1969, 1832.

12) R.M. Dell, F.S. Stone, and P.F. Tiley, Trans. Farad. Soc., 49, 201 (1953).

13) C.R. Adams, in Proc. 3rd Int. Cong. Catalysis, Amsterdam, 1964, Vol. I, p.240, North-Holland Publ., Amsterdam, 1965.

On the Mechanism of Oxygen Activation during Carbon Monoxide Catalytic Oxidation on Silica-Gel Supported Vanadium Pentoxide

M. J. Kon, V. A. Shvets and V. B. Kazansky

Institute of Organic Chemistry, Academy of Sciences of the USSR, Moscow

The stepwise mechanism involving successive reduction of an oxide surface with a substrate and its reoxidation with gaseous oxygen has been established for a large number of oxides. The evidence for it was provided by the coincidence between the rate of oxide surface reduction and reoxidation with that of a catalytic reaction under stationary conditions.[1] Although this scheme seems to be quite valid, it does not answer the question about the forms of oxide oxygen involved in the elementary steps of surface reduction and reoxidation.

Indeed, in the case of the oxidation of carbon monoxide, for example, electronically excited oxygen can react with CO molecules during reduction. Similarly the reoxidation of the reduced surface may also involve several elementary reactions with participation of various species, for instance O_2^-, O^-, etc. The present study has been undertaken to elucidate the role of the adsorbed radicals, O_2^- and O^-, at different stages of catalytic oxidation of carbon monoxide on silicagel supported vanadium pentoxide. The choice of this catalyst was due to oxygen adsorbed radicals being found recently on its surface by ESR.[2]

Two series of experiments were carried out. The first involved the study of the formation of the adsorbed radicals O_2^- and O^- during reoxidation of a prereduced catalyst, and their interaction with carbon monoxide at room temperature. The second involved the study of CO catalytic oxidation at high temperature on prereduced and preoxidized catalysts. The rate of the catalytic reaction has been compared with that of reoxidation of a prereduced catalyst.

The low temperature experiments were carried out in a specially designed unit. This allowed both the recording of the ESR spectra of the adsorbed oxygen free radicals and chromatographic analysis of the carbon dioxide resulting from their interaction with CO. The catalyst was prepared by impregnation of SiO_2 with a solution of NH_4VO_3, and contained 4 weight % of vanadium pentoxide. It was pre-reduced in a carbon monoxide flow at a temperature of 650°C. Then it was cooled to room temperature and, without expo-sure to air, introduced into the ESR cavity. The carbon dioxide resulting fro n the reaction of the adsorbed oxygen radicals with CO was displaced from the reactor by a helium flow and analyzed chromatographically.

It was shown[2] that tetrahedrally coordinated V^{4+} ions arise on the surface of reduced V_2O_5 / SiO_2 catalysts. After oxygen adsorption at room temperature they can form sur-face complexes containing O_2^- and O^- radicals in the coordi-nation sphere of vanadium ions. Figure 1 shows the ESR spectrum arising after oxygen adsorption at room tempera-ture and a pressure of 4 torr on a prereduced catalyst. The spectrum is due to overlapping of two ESR signals from the adsorbed O^- and O_2^- anion-radicals. The measurement of the spectra intensity shows that the amount of resulting O^- species is $6 \times 10^{17} \pm 20\%$ per gram of catalyst.

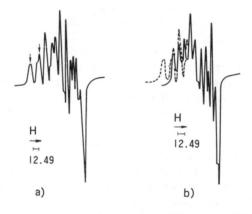

a)   b)

Fig. 1.   ESR spectra of adsorbed oxygen an iron-radicals O_2^- and O^- (the components of the O^- signal are indicated with arrows).

The ESR spectra pattern is not affected by oxygen pump-ing off. However, if carbon monoxide is introduced into the reactor after oxygen adsorption the signal from the O^- radi-cals disappears completely, whereas the spectrum from O_2^-

radicals remains unchanged, and is observed in its pure form. Chromatographic analysis shows that the amount of carbon dioxide simultaneously formed is practically the same as that of O^- which disappears. Thus, at room temperature, CO_2 would result due to the reaction of carbon monoxide with adsorbed O^-, whereas adsorbed O_2^- would not react. The scheme of the reaction can be presented as follows

$$V_{tetr}^{4+} + O_{2\ gas} \longrightarrow [V^{5+}O_2^-] \tag{1}$$

$$V^{5+}O_2^- + V^{4+} \longrightarrow 2\,[V^{5+}O^-] \tag{2}$$

$$V^{5+}O^- + CO \longrightarrow CO_2 + V_{tetr}^{4+} \tag{3}$$

Catalytic oxidation of carbon monoxide would not proceed at room temperature since quadrivalent vanadium ions regenerated in the third reaction are likely to form nonreactive complexes $[V^{5+}O_2^-]$. Nevertheless, the fact that the amount of resulting carbon dioxide is 2-3 times as high when the adsorbed radicals interact not with pure CO but with its mixture with oxygen indicates that a partial cyclic reaction may take place. Thus at room temperature V^{4+} ions can participate in carbon monoxide oxidation only 2-3 times, than the catalytic reaction becomes inhibited. It can proceed however at a temperature higher then 400°C.

The catalytic activity at a temperature of 460°C was studied in a continuous flow unit at atmospheric pressure. The resulting carbon dioxide was analyzed chromatographically. The mixture composition was 22 vol % of CO, 11% of O_2 and 67% of helium. The gas flow rate was 20 cm^3/min. The catalyst volume was 1.7 cm^3, and its weight was 1.6 g.

The activity of both the oxidized and reduced forms of the catalyst has been studied. The oxidized form was prepared by heating of the samples for one hour in a flow of oxygen at 460°C. The reduced one was prepared by heating in a flow of carbon monoxide under the same conditions. The degree of catalyst reduction was estimated from the amount of V^{4+} ions contained in it, which was determined by ESR after water adsorption. This procedure transforms tetrahedral V^{4+} to vanadyl complexes, which could be quantatively detected by ESR at room temperature.[4] The reoxidation rate of the prereduced samples was measured in the adsorption unit by oxygen consumption.

Figure 2 shows the data obtained when studying the effect of pretreatment on the activity of the catalyst, which contained 1 wt % of vanadium pentoxide. The activity of the reduced catalyst (Fig. 2, curve 1) is about 4 times higher than the stationary activity, which is the same as the activity of the oxidized catalyst (Fig. 2, curve 2). As compared with the catalyst stationary state, the amount of quadrivalent vanadium on reduction is 10 times as high, and is about 2% of total vanadium (2 μ mole/g). Figure 3 shows a kinetic curve of the oxidation at 75 torr O_2 of the prereduced sample. Its rate decreases rapidly with time. At the fourth minute it was $6 \times 10^{-2} \mu$ mole/min/g and at the twenty-first minute $5 \times 10^{-3} \mu$ mole/min/g. The reoxidation at the 460°C reaches completion about 40 min, i. e., this time is the same as that of decreasing the catalytic activity of the prereduced catalyst to its stationary value.

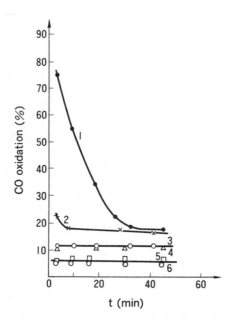

Fig. 2.   The change of activity of V_2O_5/SiO_2 catalysts.
Curve 1, reduced 1% V_2O_5/SiO_2; 2 - oxidized 1% V_2O_5/SiO_2;
3 - reduced 5% V_2O_5/SiO_2; 4 - oxidized 5% V_2O_5/SiO_2;
5 - reduced pure V_2O_5; 6 - oxidized pure V_2O_5.

By comparing the reaction rate (6 mole CO_2 per gram of catalyst per minute) with that of catalyst reoxidation under stationary conditions of reaction (V^{4+} content, 0.2 μ mole/g; reaction rate, $5 \times 10^{-3}\mu$ mole/g/min) it has been shown that the former value is three orders of magnitude higher than the latter. Thus, the subsequent reduction-oxidation mechanism is invalid for catalyst containing 1% vanadium pentoxide. Other evidence for a low rate of reoxidation as compared with that of the catalytic reaction is provided by the fact that the activity of the prereduced catalyst is practically constant after ten minute oxidation in air.

Roiter and Usa[3] and Boreskov et al.[1] have previously come to a conclusion about a "nonstaged" mechanism of hydrogen oxidation on unsupported vanadium pentoxide. The discrepancy between the rate of the catalytic reaction and that of the catalyst redox stages observed by the authors was not greater than several times. The difference between these values and those found in the present study is at least a hundred times. In our opinion, this is due to the specificity of the mechanism of oxidation on a supported catalyst with a low V_2O_5 content.

The ESR spectra of V^{4+} ions in low vanadium pentoxide catalysts shows hyperfine structure typical of isolated ions.[2] With increase of the supported oxide content up to 10%, a separate V_2O_5 phase is likely to arise. All this allows us to conclude that CO oxidation on a catalyst containing 1% V_2O_5 is most likely to proceed with participation of isolated vanadium ions as active centers. This conclusion is also

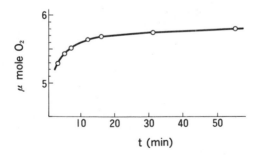

Fig. 3.   The kinetics of reoxidation of pre-reduced 1% V_2O_5/SiO_2 catalysts at 460°C and 75 torr of oxygen.

supported by the dependence of the stationary catalytic activity of the supported catalysts on their vanadium pentoxide
content, as shown in Figure 4. With decrease of vanadium
concentration this passes a maximum which corresponds to
about 1% V_2O_5, the activity with respect to one vanadium ion
increasing quite markedly. Thus, the stationary specific
activity of a sample containing 1% V_2O_5 is about 20 times
that of a sample containing 10% vanadium pentoxide. For
the prereduced samples this difference is several times
higher. It must be noted that at concentrations of 5 wt %
V_2O_5 and higher, the effect of increase in catalytic activity
after prereduction was not observed (Fig. 2, curves 3 and
4). Probably this phenomena may be explained by the formation of a separate V_2O_5 phase, whose activity also does
not change after prereduction (Fig. 2, curves 5 and 6).

It seems probable that the "nonstaged" mechanism of
reaction at high temperature can be accounted for by the
elementary reactions of the oxygen adsorbed radicals pre-
posed above for reaction at room temperature. The overall
mechanism of high temperature carbon monoxide oxidation

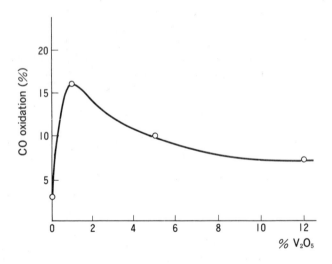

Fig. 4.   The dependence of catalytic activity of V_2O_5/SiO_2
on the amount of supported V_2O_5.

can be represented by the following chain scheme:

$$O^{2-} + CO \longrightarrow CO_2 + 2e \qquad \text{catalyst} \qquad (1)$$

$$2e + 2V^{5+} \longrightarrow 2\ V^{4+} \qquad \text{reduction} \qquad (2)$$

$$V^{4+} + O_2 \longrightarrow [V^{5+}O_2^-] \qquad \text{catalyst} \qquad (3)$$

$$[V^{5+}O_2^-] + V^{4+} \longrightarrow 2[V^{5+}O^-] \qquad \text{catalyst} \qquad (4)$$

$$[V^{5+}O_2^-] + V^{4+} \longrightarrow 2\ V^{5+} + O^{2-} \qquad \text{reoxidation} \qquad (5)$$

$$[V^{5+}O_2^-] + CO \longrightarrow CO_2 + [V^{5+}O^-]\ \text{reaction} \qquad (6)$$

$$[V^{5+}O^-] + CO \longrightarrow CO_2 + V^{4+} \qquad \text{reaction} \qquad (7)$$

If the rate of reaction of the oxygen adsorbed radicals with carbon monoxide (6 and 7) is much higher than the rate of catalyst reoxidation (4 and 5) then cyclic repetition of stages 3, 6, 7 and 4, 7 becomes possible. As a result the rate of the catalytic reaction may considerably exceed the rate of surface reoxidation.

Reactions 4 and 5 involve additional ions of quadrivalent vanadium and electron transfer along the catalyst lattice. With an unsupported oxide which is a semiconductor, the rate of the process is high and the rate of the reoxidation stages approaches the overall rate of the catalytic reaction. In a dilute supported catalyst, where electron transfer between the quadrivalent vanadium ions is hindered because of the insulating properties of the support, the rate of the reoxidation stage sharply decreases. As a result, for each elementary act of V^{4+} reoxidation there are several hundreds of elementary reactions of the oxygen adsorbed radicals with carbon monoxide, and "chain" reaction takes place. It seems probable that the "nonstaged" mechanism of catalytic oxidation is common for supported catalysts containing quasi-isolated ions of transition metals.

References

1) G. K. Boreskov, Kinetika i Kataliz, 2, 374 (1970).
2) V.A. Shvetz, V.M. Vorontinsev, and V.B. Kazansky, Kinetika i Kataliz, 10, 365 (1969).
3) V.A. Royter and V.A. Usa, Kinetika i Kataliz, 3,

343 (1962).

4) L. L. Van-Reijen and P. Cosse, Disc. Farad. Soc.,
    41, 277 (1966).

Discussion

A. P. Purmal

I would like to comment on the mechanism proposed by Dr. Moro-oka for the catalytic oxidation of propene to acetone on a mixed molybdenum-tin oxide catalyst. The stage of H_2O conversion into H and OH on the surface under the experimental conditions used by Dr. Moro-oka seems unlikely. I think that from chemical and theoretical points of view the following mechanism which is in good agreement with the main experimental results will be more attractive.

$$C_3H_6 \longrightarrow (C_3H_6)_S \qquad (1)$$

$$H_2O \longrightarrow (H_2O)_S \qquad (2)$$

$$(H_2O) \rightleftharpoons (H^+)_S + (OH^-)_S \text{ but not } (\overset{\bullet}{H}) + (\overset{\bullet}{O}H) \qquad (3)$$

$$(C_3H_6)_S + (H)_S \longrightarrow (C_3H_7^+)_S \qquad (4)$$

$$(C_3H_7^+)_S + (OH^-)_S \longrightarrow (C_3H_7OH)_S \qquad (5)$$

For example, the stage of propene hydration where the introduction of ^{18}O from $H_2{}^{18}O$ into oxidation products just takes place. But according to the proposed mechanism this introduction takes place as a result of acid-base catalysis as opposed to the mechanism which Dr. Moro-oka proposed for redox catalysis ($\overset{\bullet}{O}H + C_3H_7 \longrightarrow C_3H_7OH^+$). The conversion of C_3H_7OH to acetone can be shown with the help of the data reported by Dr. V. B. Kazanski on the role of O^- in the process of heterogeneous redox catalysis. The scheme can be expressed as follows

$$Mo^{5+} + O_2 \longrightarrow Mo^{6+} + (O_2^-)s \qquad (6)$$

$$(\overset{\bullet}{O_2})s + Mo^{5+} \longrightarrow Mo^{6+} + (2\overset{\bullet}{O}^-)s \qquad (7)$$

$$(\overset{\bullet}{O}^-) + (H^+) \rightleftharpoons (\overset{\bullet}{O}H)s \qquad (8)$$

$$(O\overset{\bullet}{H})s + (C_3H_7OH)s \longrightarrow H_2O + (C_3H_6^{\bullet}OH)s - \alpha \qquad (9)$$

-alcohol radical

$$(C_3H_6OH)s \longrightarrow CH_3COCH_3 + (\overset{\bullet}{H})s \qquad (10)$$

$$(\overset{\bullet}{H})s + Mo^{6+} \longrightarrow (H^+)s + Mo^{5+}. \qquad (11)$$

Y. Moro-oka

How can it be explained on the basis of the mechanism proposed by you that the catalytic activity of samples depends on the acidic properties of the catalyst surface?

A. P. Purmal

It is necessary to assume that the stage of the formation of $C_3H_7^+$ is rate determining. It is quite possible, as the other reaction (9) which might be considered as a rate determining step seems to occur very rapidly. In a solution the rate constant K for this reaction is about 10^9 l/mole. sec.

V. D. Sokolovski

I join in the high evaluation of Dr. Moro-oka's work given in the preceding speeches. But I have some remarks in relation to unambiguous conclusions of this work.

The main conclusion of the author on the immediate participation of water in the process of oxidation is based on two facts: the first, low rate of oxygen isotope ^{18}O exchange between water and gas phase oxygen; the second, different content of ^{18}O isotope in acetone, acrolein, and acetaldehyde obtained simultaneously on Sn-Mo catalyst at high temperatures. This makes it possible for the author to conclude that the appearance of ^{18}O in acetone results from the direct interaction of propylene with water enriched with this isotope. But it is quite possible that the difference in the content of ^{18}O isotope in acetone and other products observed by the author may result from the exchange of acetone with water. From the author's data the rate of the exchange between acetone and water even at lower tompera- tures provides the ^{18}O isotope content found in acetone.

Thus, the data obtained do not permit us to reject the following scheme of ^{18}O introduction to products: rapid isotope exchange between water and an oxide followed by the oxidation of olefin by catalyst oxygen.

The fact that the appearance of ^{18}O isotope in reaction products is observed only at low temperatures (with the exception of acetone which can be enriched at the expense of exchange) also merits consideration. It can be explained if it is remembered that at the evaluation of the difference in activation energies of the exchange and reaction at high temperatures the rate of a reaction can already exceed that of the exchange.

In conclusion I want to emphasize that I do not reject the mechanism proposed by the author but believe that additional data are needed for its support.

G. I. Golodets

Two years ago we published a paper (Doklady Akademii Nauk USSR, 1969) in which two main criteria of an optimum catalyst for reactions of partial oxidation of organic compounds to acids or their anhydrides were formulated: 1) moderate values of the bond of oxygen with an oxide surface; 2) amphoteric properties of a catalyst surface. Only the combination of these two properties gives a good catalyst for incomplete oxidation. More details on this problem were given in our report in this seminar.

The reported works by Professor Moro-oka and Professor Seiyama show that the acid-basic properties of catalysts should be taken into account for other reactions of incomplete oxidation (oxidation of olefins to cations, etc.) as well. Thus the concept of an important role of acid-basic properties (along with the stability of a surface oxygen bond) which we advanced for reactions of organic acid formation seems to be of fairly general importance for reactions of incomplete oxidation. It is natural that the optimum value of acid-basic properties should depend on the type of an incomplete oxidation reaction, as the structure and properties of intermediate compounds and activated complexes for different reactions are unequal.

T. Kwan

Professor Kazanski's report involves some interesting problems of heterogeneous catalysis: 1) the nature of supported and unsupported catalysts and 2) the molecular description of a so-called "reduction-oxidation mechanism". Let us assume according to the report that Stages 3, 6 and 7 take place. Then in this cyclic process Stage 3 will be

rate determining. With such a mechanism the "reduction"
of a catalyst is still important for the determination of
catalytic activity.

In connection with the report by Professor Boreskov I
would like to make the following remark. The attempts to
understand catalytic activity on the basis of the Brönsted-
Polanyi-Semenov rule are crude but useful, in particular
while searching for catalysts. Nevertheless, I belive that
the treatment by this method should further be supported
by a molecular description.

Finally, I have some words on the interpretation of a
kinetic isotopic effect. The adsorption of H_2 on ZnO is
known to form H-Zn and OH(IR) bands. If the interaction
of hydrogen of H-Zn with oxygen of a hydroxyl is assumed
to be the slowest stage of oxidation of H_2, then the observed
kinetic isotopic effect can be reasonably interpreted. The
zero-point energy difference in H-Zn and D-Zn bonds is
entirely consistent with the observed value of K_H/K_D, as
well as with its temperature coefficient (IV International
Congress on Catalysis, Moscow, 1968).

I. I. Tretyakov
In the report by Professor G. K. Boreskov the influence
of the energy of the bond of products with a catalyst surface
on a reaction rate was considered in great detail, and for
the first time the availability of optimum values for this bond
energy was shown. It was interesting to estimate the
maximum value of the energy of the reaction products-
catalyst bond at very rapid reactions observed on metals
cleaned in superhigh vacuum. Thus, while oxidizing carbon
oxide by oxygen on clean platinum or palladium, the reaction
rate at CO pressure of 10^{-2} torr under optimum conditions
when the surface is completely covered with oxygen is equal
to 5×10^{17} mole/cm^2 sec. In a nonstage catalytic process
the same quantity of molecules of a reaction product (i. e.,
CO_2) should have time to desorb from the surface. Using
the Wagner-Polani equation, it can be found that in this case
the maximum value of the energy of the bond of reaction
products with a surface is only several kcal/mole. It seems
to be one of those cases considered in the report when the
energy of the products-catalyst bond does not affect the
activation energy of the chemical conversion stage.

L. Ya. Margolis

The data on adsorption and the results of studying
gamma-resonance spectra work functions show that on the
surface of oxide catalysts hydrocarbon-oxygen complexes
are actually formed. We showed earlier on various catalysts
of deep and partial oxidation that these complexes have different composition and opposite charges. The use of
gamma-resonance spectroscopy made it possible not only to
find out the formation of complexes on the surface of catalysts containing tin or iron ions but also to determine the
change of these ions.

The comparison of tin dioxide, representing a complete
oxidation catalyst with a tin-molybdenum catalyst for partial
oxidation, and the analogous comparison of ferric oxide with
an iron-molybdenum catalyst showed that only in the presence of a molybdenum ion capable of forming a complex with
olefin does the electron transfer resulting in change of the
valence of tin and iron ions take place. We believe that a
similar transfer should also take place in the case of cobalt-
molybdenum and bismuth-molybdenum catalysts. Oxygen-
hydrocarbon complexes performing partial oxidation seem to
be binuclear and connected with lower-valent ions (bivalent
ions of tin, iron, etc.) as well as with a molybdenum ion.
On the contrary, complete oxidation will require steady
adsorption of olefin on one active center (a cation or a
cation-oxygen pair) as well as sufficiently mobile oxygen
capable of reacting with an olefin molecule without breaking
its bond with a surface but performing a certain surface
chain reaction. Therefore they need mobile oxygen with low
bond energy.

Electron transfers and valence change in catalysts can be
enhanced by the appropriate set of ions with a certain electron structure. That is why partial-oxidation catalysts
become multicomponent (involving three or four components). In this connection it is interesting to reconsider the
mechanism of the effect of impurities on the selectivity of
partial hydrocarbon oxidation. Assuming that in complex
oxide systems the localization of electrons on single ions or
groups of ions occurs, impurities do not practically change
the electronic properties of such modified samples (small
change of the electron work function for bismuthmolybdate,
etc.). The impurities affect: 1) the environment of the ion
being a catalysis active center and adsorption; 2) the cata-

lyst structure promoting the appearance of most active
phases; and 3) the oxygen concentration on the surface (in
the case of electronegative impurities).

The ion environment can be estimated by the change of
quadrupole splitting, which characterizes the gradient of an
electric field around a cation induced by the disturbance of
the charge distribution symmetry in a solid lattice which
does not depend on the nature of bonds; the structure of the
compound, and the electronegativity of adjacent atoms. For
partial oxidation, impurities enhance selectivity, provided
that the quadrupole splitting increases as compared with
reference samples.

In the example of bismuthmolybdate the change of the
structure was shown during the modification by various
admixtures containing elements of different electron structures. The most active and selective catalysts are β-phase.

The addition of electronegative admixtures to a gas, as it
was shown earlier, can also be used to control oxygen
concentration on the catalyst surface. In the recent work by
Kallis and Keene it was possible to oxidize methane to
formaldehyde on palladium capable only of converting this
molecule to carbon dioxide and water.

S. L. Kiperman

Professor Seiyama's report dedicated to the comparison
of catalytic activity and selectivity on different catalysts was
of great interest. The precise measure of a catalytic
activity value can be the values of a reaction rate or other
values, the change of which while passing from one catalyst
to another are symbatic to the values of a reaction rate. As
Professor Seiyama made his experiment in a flow system,
he could not measure a reaction rate. Therefore Professor
Seiyama uses for the comparison either the degrees of conversion at constant contact time or contact time at a constant
degree of conversion. But the changes of these values can
be symbatic to the changes of a reaction rate only under the
condition when the kinetics does not change. As there are no
data on the kinetics of a reaction, the use of the characteristics of activity and selectivity for comparison might insufficiently precise.

V. N. Berdnikov

During the adsorption of molecular oxygen on partially
reduced oxides, i. e., on TiO_2 or V_2O_5 ion radicals, O_1^- or

O_2^- are formed, being present on the surface in the form of complexes of $V^{5+}O_2^-$ and $V^{5+}O^-$ type. In Professor Kazanski's report the assumption was made of a possible role of these particles in heterocatalytic oxidation of CO to CO_2. It is interesting to note that the analogous complexes of the O_2^- ion radical are also formed in a liquid phase in some oxidation-reduction systems. Thus, we observed the spectra of EPR complexes of O_2 with ions Ti (IV), Zr (IV), Hf (IV), Th (IV), V (V), and Nb (V) while performing under flow conditions the reaction between Cl (IV) and H_2O_2 in the presence of the above-mentioned ions of metals. The EPR spectra of these complexes at 77°K resemble the analogous spectra of the O_2^- ion radical on the proper oxides (in the case of V_2O_5 and TiO_2). The ions Ti (IV), Zr (IV), etc., are known to catalyze various oxidation-reduction reactions in solutions, e.g., the oxidation of iodide or thiosulfate by hydrogen peroxide. It is possible that complexes with O_2 play a certain role in the mechanism of these reactions. This example shows that as regards the elementary act the problems of homogeneous catalysis of processes of oxidation by molecular oxygen have a great deal in common.

I would like to discuss the reaction of oxidation of CO on V_2O_5 from the energetic point of view and try to evaluate the energy of the formation of $V_n^{5+}O_2^-$ and $V_n^{5+}O_2$ surface complexes. Let us consider the reaction given in V. B. Kazanski's report:

(I)

$$CO\ g + O^-g \longrightarrow CO_2g + e^-g \qquad \Delta H_0 = -93\ kcal$$

$$V_n^{5+}O + CO^-g \longrightarrow V_n^{4+} + CO_2 \qquad \Delta H_1 = -93\ kcal + \Delta Q_0 = I$$

(II)

$$CO\ g + O_2^-g \longrightarrow CO_2g + O^-g \qquad \Delta H_0 = -22\ kcal$$

$$V_n^{5+}O_2^- + CO\ g \longrightarrow V_n^{5+}O^- + CO_2g \qquad \Delta H_2 = -22\ kcal + \Delta Q_{0_2} = \Delta Q_0,$$

where I is the potential of ionization of surface V_n^{1+}, ΔQ_{0_2} the heat of complex formation of V_n^{5+} with O^-, ΔQ_{02} the analogous value for O_2^-, and ΔH_0, ΔH_1, ΔH_2 the enthalpies. The calculation of heat of gas reactions is made on the basis of table thermodynamic data.

Now let us consider the process:

(III)

$$V_n^{4+} + O_2g \longrightarrow V_n^{5+}O_2^- \qquad \Delta H_3 = -20,8 + I - \Delta Q_{O_2^-} ,$$

where 20.8 is the affinity to electron for O_2. According to
V. B. Kazanski's data, when oxygen pressure decreases, the
signals (EPR) of $V^{5+}O_2^-$ complexes reversibly disappear.
That means that the change of free energy in Reaction (III)
is equal to a small negative value of about several kcal. As
the changes of entropy in Reaction (III) primarily depend on
the loss of the translation entropy of the O_2 molecule, i.e.,
is about 36 e.u., we obtain the following approximate
estimation:

$$\Delta H_3 \simeq -15 \text{ kcal}$$

so that:

$$I - \Delta Q_{O_2^-} \simeq 6$$

The potential of ionization I should be equal to several
eV; and as far as one can judge by the available data for
other oxides it can be hardly less than 4 eV. Let $\tau = 4$ eV.
Then $\Delta Q_{O_2^-} \simeq 90$ kcal. According to the data presented in
V. B. Kazanski's report, the $V_n^{5+}O_2^-$ complex does not react
with CO at room temperature, but it seems to react with it
at temperatures of 400-500°C. Therefore, it is quite
reasonable to assume that Reaction (III) is endothermic and
the value ΔH_2 is equal to at least +10 kcal. From this it
follows:

$$\Delta Q_{O_2^-} - \Delta Q_{O^-} \simeq 30 \text{ kcal}$$

Thus

$$\Delta Q_{O^-} \simeq 60 \text{ kcal.}$$

Then for ΔH_1 we obtain

$$\Delta H_1 \simeq -125 \text{ kcal.}$$

It should be expected that this reaction should easily occur even at decreased temperatures, which agrees well with the data given by V. B. Kazanski.

From the evidence of the evaluations performed it is clear that the found heat values of the formation of $V_n^{5+}O_2^-$ and $V_n^{5+}O^-$ complexes represent approximate evaluations of lower limits of these values. It is evident that for a more precise evaluation of these values the direct measurements of ΔH_3 and I are of interest. But true values of $\Delta Q_{O_2^-}$ and ΔQ_{O^-} are unlikely to be significantly higher than the evaluations obtained by us.

I. S. Sazonova

In the communication by V. I. Marshneva, G. K. Boreskov, and V. D. Sokolovski the mechanism of CO oxidation on nickel oxide and titanium dioxide is discussed. At the laboratory of semiconductive catalysts of the Institute of Catalysis, N. P. Keier, I. S. Sazonova, I. L. Mikhailova, A. E. Cherkashin, and P. V. Bunina also investigated this reaction on TiO_2 (rutile) and NiO. In contrast to the work by Marshneva et al. we investigated the reaction under nonstationary conditions on catalysts preliminary treated in vacuum. On the basis of the study of reaction kinetics and reactant chemisorption, the direct comparison of rates and energies of activation of processes of carbon oxide oxidation and catalyst reduction by carbon oxide, as well as in the set of experiments with ^{14}CO, we found that both on TiO_2 (up to 600°C) and NiO (up to 450°C) a reduction-oxidation mechanism does not exist. The reaction rate-determining stage is the interaction of CO with chemisorbed oxygen. In this respect our data are consistent with the results of the work reported.

But the investigation of the catalytic and electric properties of titanium and nickel oxides with additives along with their clean oxides permitted us to find out the influence of an electron factor on the process of CO oxidation. For titanium dioxide alloyed by tungsten, uranium, tantalum, antimony, and iron, as well as for nickel oxide alloyed by lithium, the linear relationship between the changes of the reaction activation energy and the electron work function was found. The additions of lithium to NiO and of tungsten, tantalum, and antimony to TiO_2 result in an increase of the work function and reaction activation energy. The additions of uranium and iron to TiO_2, as well as incomplete reduction of titanium dioxide to $TiO_{1.96}$, decrease the work function and reaction activation energy. We believe the fact that the character of the relationship between the activation energy and work function is the same both on alloyed samples of nickel oxide with hole-type conductivity and on alloyed samples of titanium dioxide with electron conductivity is of interest.

The presence of this relationship indicates that the formation of a transition complex on a limiting stage occurs with the participation of free electrons of a catalyst. So the limiting stage is acceptor and can be written in the following way:

$$CO \; g + O^-(ads) + e^- \longrightarrow CO_2^{2-}(ads).$$

Kazanski et al., while studying CO oxidation on TiO_2 (anatase) by the EPR method, also showed the participation of electrons of conductivity in the limiting stage.

O. V. Krylov

I would like to make one remark on the communication by Berdnikov. After our calorimetric measurements the heat of oxygen adsorption on Mo and V catalysts with the formation of O_2^- is about 20 kcal/mole. It seems to me this value does not agree with Berdnikov's calculations.

V. B. Kazanski

In the report by Marshneva and Boreskov the conclusion was made on the nonstage type of the mechanism of carbon oxide oxidation on vanadium pentoxide, supported and non-supported onto silica gel, which is in agreement with our conclusion. Nevertheless, I would like to emphasize the difference between the systems studied. We worked with a catalyst containing only 1% vanadium and found the increase of a reaction rate at sample reduction and approximately one hundred times higher difference in the rates of a catalytic reaction and sample reoxidation than in the work of G. K. Boreskov. We related these results with a specific feature of a diluted system containing quasi-isolated vanadium ions.

In the work of Marshneva and Boreskov at sample reduction the reaction rate decreased and the difference between deposited and non-deposited systems was not found. We believe that it is explained by the fact that the authors studied

a supported catalyst having too high vanadium content (10%)
that resulted in disappearance of the supported sample
specific feature. Indeed, the available additional results
show that while increasing vanadium pentoxide content in
samples the increased activity connected with sample prereduction disappears.

On Professor Boreskov's report I would like to note the
following: Owing to the works of Professor Boreskov's
school, the importance of correlations between oxygen bond
energy on the oxide surface and catalytic activity became
evident. Nevertheless, the physical significance of these
correlations is not completely clear, since the active form
of reacting oxygen (oxygen radicals, variable-activated
oxygen, etc.) is unknown. The investigation of the oxygen
activation mechanism would make it possible to understand
the correlation between oxygen bond energy and catalytic
activity a more profoundly.

V. D. Sokolovski

I would like to emphasize one of the results of our work in
connection with the report by V. B. Kazanski.

While investigating the relationship between stage and
associative mechanisms of carbon oxide oxidation on V_2O_5,
we studied the relationship between a reaction rate and V^{4+}
concentration in a catalyst. It was proved that in stationary
samples at different temperatures the quantity of V^{4+} directly
measured changes very slightly while the reaction rate
differs by orders. What is more, the reaction rate at constant temperature on samples with different V^{4+} content
decreases with the increase of the oxide reduction degree.
This contradicts the scheme of carbon oxide oxidation offered
by V. B. Kazanski for a supported vanadium catalyst. The
centers of carbon oxide oxidation on vanadium pentoxide
following the nonstage mechanism seem to be not V^{4+} but
some other local activations appearing in the process of
oxide oxidation-reduction.

G. I. Golodets

I would like to direct a question to Professor Moro-oka:
How can you explain the positive effect of water vapors on
selectivity of catalysts in partial oxidation of propylene to
acrolein?

Y. Moro-oka

There are two cases:

The first case is the oxidation of propylene to acrolein on Pd-activated charcoal catalyst. In this case oxygen to acrolein comes from water and not from molecular oxygen. Here the positive effect of water is clear.

The second case is the oxidation of propylene on SnO_2-MoO_3 or Bi_2O_3-MoO catalyst. In this case the presence of water vapors is known to increase the selectivity of acrolein although an oxygen atom comes from molecular oxygen (or metal oxide oxygen).

I have no evidence supporting the answer. But two possible cases can be considered to explain the positive effect of water on the selectivity of acrolein:

1. The adsorption of water vapors changes the work function of a catalyst. Consequently, the adsorbed states of oxygen and propylene change.

2. The adsorption of water takes place on the active center with the formation of carbon oxides.

V. V. Popovski

We are grateful to Professor T. Kwan for the interest in our work. Certainly, we know the works of Japanese scientists and, in particular, the work on the determination of hydrogen isotopic effect in hydrogen oxidation on zinc oxide. But for lack of time I could not even to mention the works preceding ours.

Now I shall answer particular remarks on the report:

1. We agree with Professor T. Kwan that a difficult stage of hydrogen oxidation for the most of the investigated oxides is the interaction of oxides with hydrogen. But it should be kept in mind that under the conditions of a stationary regime the rates of reduction and oxidation stages are the same. As such a catalytic system is self-regulated, the nature of a difficult stage will determine only the stationary degree of the catalyst surface reducibility. Under the conditions of our experiments the degree of surface oxidation was high (98-99% of oxygen surface monolayer), and this permits us to draw a conclusion about the stage of hydrogen interaction with oxides as being difficult.

2. We eliminated the possibility of surface hydride formation as intermediate compounds at catalytic hydrogen oxidation on chromium oxide for two reasons. First, if hydrogen is dissociatively adsorbed on chromium atoms,

hydrogen atoms could recombine and be isolated into a gas phase. It would result in hydrogen isotopic exchange. But under the condition of hydrogen oxidation on chromium oxide the exchange was absent. Secondly, as we observed, hydrogen is irreversibly adsorbed on chromium oxide and is isolated into a gas phase in the form of water. In Trapnell's opinion this points to hydrogen adsorption on oxygen surface atoms as adsorption centers.

G. K. Boreskov

I would like to clarify the problem of the stage mechanism. The measurements given make it possible to consider it on the basis of precise quantitative results. It has been shown that for a number of catalysts under certain conditions and temperatures the rates of catalyst oxidation and reduction stages coincide with the rate of a catalytic process. It does not eliminate the fact that under other conditions for the same catalysts the process can follow an associative (non-stage) mechanism. It seems to me that it is a widespread rule: at low temperatures it is an associative mechanism, and at high temperatures, a stage one. It can be often seen on Arrenius curves when two regions are observed that the high temperature region of the reaction course with reasonably high activation energy corresponds to a stage mechanism and the low temperature region with low activation energy and small enthropy of an active component shows the reaction course by an associative mechanism. In this case it is possible that the stage mechanism, as was rightly shown by Professor Kwan, includes complex stages summarizing the number of elementary processes.

I agree with Professor Kazanski that a more profound study of the mechanism is necessary. I do not like the expression "an activated form of oxygen" in relation to reacting oxygen. It can lead to misunderstanding, as it is associated with the appearance of some energy-enriched forms of oxygen in superequilibrium quantitatives. As a rule it does not take place. It is possible for the reaction to follow a chain mechanism with nonequilibrium concentrations, but it is a rare case of catalysis. As a rule the acceleration by a catalyst is achieved by increasing the number of particles with high energy on the surface but as a result of the realization of a new reaction route which does not require high activation energy.

During oxidation of H_2, CO and during the analogous reactions the breaking away of oxygen is a limiting stage. This was proved quantitatively by the measurements of the stage rates, and furthermore, it directly follows from the correlation between the energy of oxygen bond and catalytic activity.

Unfortunately, in relation to kinetic isotopic exchange V.V. Popovski did not give quantitative data. I want to emphasize the main result of the work: qualitatively found kinetic effect is normal: a lighter isotope reacts more quickly. The small value of the isotopic effect is difficult to explain. It leads us to assume that hydrogen enters an activated complex with little deformation of the bond and possesses sufficient mobility.

In connection with Professor Sokolovski's speech concerning my report, I am glad that his data qualitatively supported the results of my consideration of the role of the energy of the product-catalyst bond. However, I should note that the increase of the reaction rate with increasing concentration of products observed by Professor Sokolovski cannot be explained on the basis of my concepts and needs additional studies.

Electronic Interaction of Molecular Oxygen with Metal Porphyrins in Solution

K. YAMAMOTO and T. KWAN

University of Tokyo, Tokyo

Iron porphyrin complexes are well known to take part in oxygen-carrying electron transfer (redox reactions) and decomposition of hydrogen peroxide in biological processes. These functions may arise from both the iron porphyrin and the axial ligand environment(protein), and are quite informative in regard to the catalytic mechanism.

In the present work a much simpler reaction system is condisered. The interaction of synthetic iron tetraphenylporphyrin (abbreviated to TPP) complexes (Fig. 1) with molecular oxygen has been investigated in chloroform solutions, mainly by means of an ESR technique, with special interest in the electron-spin configurations of the central metal ion and its ligand environment. Since the spin state of iron in the porphyrin ring is dependent on the coordinating ability of the axial ligand, the redox behavior of the iron will be affected more

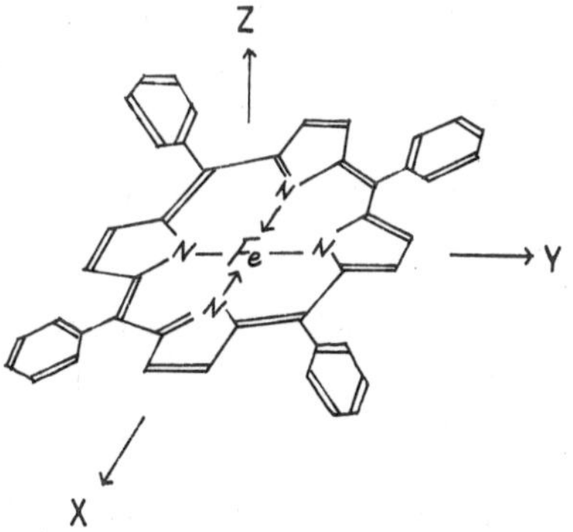

Fig. 1. Iron tetraphenylporphin. Four phenyl groups are bound perpendicular to the porphyrin plane.

73

or less by the surroundings. Previous work[1-3] on cobalt
phthalocyanine and cobalt TPP has in fact shown that oxygenation and oxidation catalysis by these complexes is extensively influenced by the presence of axial ligands.

It is the purpose of the present paper to outline such
redox behavior of the iron porphyrin complexes and to compare
it with those of manganese and cobalt TPP complexes.

EXPERIMENTAL

Fe(III)-, Mn(III)- and Co(II)-TPP complexes were prepared as usual[4] from metal(II) acetate and synthetic TPP.
These complexes were dissolved in chloroform with or
without base, which is expected to serve as axial ligand
(50-200 times as concentrated, or saturated). The axial
bases used were tetrahydrofuran, quinoline, pyridine (5.18),
4-cyanopyridine(1.80), 4-methylpyridine(6.10), 4-aminopyri-
dine(9.18), piperidine(11.12), imidazole(7.12) and benzimida-
zole(the figures in parentheses indicate the pK_a of the base).

Divalent iron and manganese complexes were prepared
by reducing the corresponding trivalent metal TPP complexes
with sodium dithionite($Na_2S_2O_4$). After the reduction treatment, the reducing reagent was removed from the solution
by centrifugation. The concentration of metal complex
ranged usually from 10^{-3} to 10^{-2}M. Throughout the procedure
the complexes were kept in a dried nitrogen atmosphere.

The oxidation of metal(II) complexes was carried out by
exposing their chloroform solutions to oxygen (one atm) at
room temperature. ESR or optical absorption spectra were
recorded before and after the exposure to oxygen. ESR
spectra were recorded at 77°K by a JEOL-P-10(X-band)
spectrometer with 100 kcps field modulation.

RESULTS AND DISCUSSION

Spin State of Iron(III) Complexes
Magnetic susceptibility measurements have served to
provide information on the electronic configuration of ferrous or ferric ions. ESR data have also proved to be supplementary to this sort of information. Thus, theoretical
analyses[5,6] have shown that high-spin iron(III) porphyrin complexes should exhibit ESR spectra of axial symmetry at $g_{\perp} \sim 6$
and $g_{\parallel} \sim 2$ while low-spin complexes should give anisotropic

spectra of three different g components near g ∼ 2. The spin
state of iron(III) porphin may thus be judged qualitatively
from the shape of the spectra obtained.

A typical ESR spectrum of Fe(III)-TPP complex in
chloroform is shown in Fig. 2. It is apparent from the
spectral shape in Fig. 2 that the complex is high-spin. The
spin configuration is consistent with that determined by
magnetic susceptibility measurement[?] On adding excess
imidazole to this solution, the spectrum of Fig. 2 appeared
to change largely into that shown in Fig. 3, or that characteristic of low-spin iron. A weak absorption around g ∼ 6
would probably be due to contamination by high-spin iron.

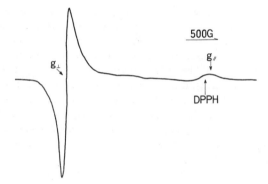

Fig. 2.   ESR spectrum of Fe(III)-TPP-Cl in chloroform at
77°K. $g_\perp \sim 6$, $g_\parallel \sim 2$.

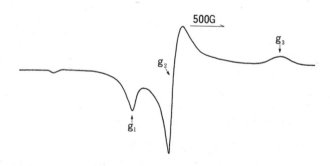

Fig. 3.   ESR spectrum of [Fe(III)-TPP-2Im]Cl in chloroform at 77°K. g = 2.98, g = 2.31, g_3 = 1.52.

A crystalline sample was then isolated after careful, repeated partial evaporation and centrifugation. Elementary analysis of the isolated crystal indicated that its composition was consistent with [Fe-TPP-2Im]Cl (imidazole is abbreviated to Im); $C_{52}H_{36}N_8FeCl$, anal. (%), calcd., C : 71.69, H : 4.06, N : 13.37; found, C : 71.29, H : 4.47, N : 14.99. The sample redissolved in chloroform gave almost the same spectrum as that shown in Fig. 3. The diimidazole complexes may hence be regarded as stable in chloroform. Its formation is expressed by

$$Fe(III)\text{-}TPP\text{-}Cl + nIm \longrightarrow [Fe(III)\text{-}TPP\text{-}2Im]Cl + (n-2)Im$$

ESR spectra of Fe(III)-TPP complexes coordinated with the other axial bases were further investigated, and most complexes were found to be high-spin, except for that coordinated with 4-aminopyridine, which was low-spin. Inspection of these spectra permitted us to confirm that a high basicity of the axial ligand favours a low-spin configuration of iron. The electronic features of the Fe(III) complexes are illustrated in Table I. It is interesting to note in the table that the dipyridine iron(III) complex is high-spin whereas the diimidazole complex is low. This rather remarkable difference between the pyridine and imidazole ligands may partly be ascribed to a stronger π-donating capability by imidazole as compared with pyridine,[8] in addition to the stronger basicity of imidazole.

Table I.   Electron Spin States of Fe(III)-TPP Complexes.

Complex	ESR spectra	Spin state
Fe(III)-TPP-Cl	$g_\perp=6.10$, $g_\parallel=1.98$	High-spin
[Fe(III)-TPP-2B]Cl		
B : 4-cyanopyridine pyridine 4-methylpyridine	$g \sim 6$	High-spin
4-aminopyridine	anisotropic	Low-spin
B : imidazole	$g_1=2.98$, $g_2=2.31$, $g_3=1.52$	Low-spin

Spin State of Iron(II) Complexes

Low-spin iron(II) complexes have six d electrons in the $t_{2g}(d_\pi)$ orbital and hence are diamagnetic. ESR absorption for high-spin iron(II) complexes is difficult to measure, probably because of a short spin-lattice relaxation time. Fe(II)-TPP coordinated with various bases has in fact shown no ESR spectrum, indicating that this technique is inadequate as far as ferrous iron is concerned.

On the other hand, magnetic susceptibility data were available for Fe(II)-TPP complexes,[7] as illustrated below.

	Bohr magneton (μ)
Fe(II)-TPP	4.75
Fe(II)-TPP-2THF	2.75
Fe(II)-TPP-2Py	0

Accordingly, the dipyridine iron(II) complex is low-spin while Fe(II)-TPP is high. Unfortunately, no information was available for the diimidazole iron(II) complex. However, it seems very probable in view of the stronger coordinating capability of imidazole as compared with pyridine that the diimidazole iron(II) complex is also low-spin.

Interaction of Oxygen with Iron(II) Complexes

The oxidative behavior of Fe(II)-TPP complexes has been investigated in the light of both ESR and optical absorption spectra before and after exposure to oxygen at room temperature. Generally speaking, the reactivity of Fe(II)-TPP complexes appeared to vary extensively from one complex to the other. Here, we shall confine ourselves only to some extreme cases.

The high-spin Fe(II)-TPP complex was readily oxidized to yield Fe(III)-TPP on exposure to oxygen, whereas for the low-spin Fe(II)-TPP-2Py this was not found to be the case. Since the latter complex is hexacoordinated, the fact that oxidation did not occur for Fe (II)-TPP-2Py would, at first sight, indicate that the axial pyridine cannot be displaced by oxygen. This interpretation was not correct however, because the hexacoordinated Fe(II)-TPP-2Im complex appeared to be readily and reversibly oxidized under exactly the same reaction conditions as in the Fe(II)-TPP-2Py case. The interesting oxidative behavior of the Fe(II)-TPP complexes is compared below with their electron spin states (in parentheses).

Fe(II)-TPP (high)   ———————→ Fe(III)-TPP (high)
Fe(II)-TPP-2Py (low)———×——→ Fe(III)-TPP-2Py (high)
Fe(II)-TPP-2Im (low)⇌Fe(III)-TPP-2Im (low)

where—→indicates "oxidized" and—×→"not oxidized". In other
words, "oxidized" indicates the formation of Fe(III) complex
when the corresponding Fe(II) complex is brought into contact with oxygen and "not oxidized" means no spectral change.

As shown above, the diimidazole Fe(II) and Fe(III) complexes were both low-spin. On the other hand, the dipyridine
Fe(II) was low-spin, with its oxidized form high-spin. The
change in the spin state of iron is remarkable for the one-
electron oxidation of the dipyridine Fe(II) complex when compared with that for the diimidazole complex (Fig. 4). According to X-ray analysis,[9] there exists a substantial difference
in structure between the low-spin iron porphyrin and high-
spin iron porphyrin complexes. For high-spin iron porphyrins the iron atom is known to lie 0.40-0.50 Å out of plane
from the four porphyrin-nitrogens whereas for low-spin iron
porphyrins it lies nearly in plane, hence bringing about
rather long bonds between the iron and the porphyrin-nitrogen
for high-spin iron porphyrins. Thus, one-electron oxidation
of the dipyridine Fe(II) complex may be forbidden according
to the Franck-Condon principle.

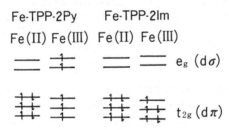

Fig. 4.   Electron configuration of Fe-TPP complexes.

The ready oxidation of the diimidazole Fe(II) complex by
molecular oxygen to yield Fe(III) should be worthy of note
because the mechanism is perhaps associated with that in
biological processes, for example, the effective electron
transfer by hexacoordinated cytochrome C. Mössbauer
studies[10] of the Fe(II)-porphyrin bond have indicated that the
electron of low-spin Fe(II) is extensively delocalized to the

porphyrin ligand. Porphyrin itself is known to undergo facile reduction and oxidation, forming a variety of molecular complexes[11,12]

It is suggested therefore that the oxidation of diimidazole Fe(II) complex by molecular oxygen takes place by a long range interaction mechanism. One electron of the $t_{2g}(d_\pi)$ orbital of Fe(II) is delocalized to the vacant π-orbital of the porphyrin ligand. Electrons may thus be transferred to molecular oxygen which is present close to the porphyrin ligand, although the position of oxygen is still left open.

Interaction of Oxygen with Manganese(II) Complexes

Mn(II)-TPP gave rise to no ESR spectrum in chloroform, but on adding an excess of imidazole the complex showed a spectrum around $g \sim 2$. The spectrum consisted of six hf lines (I =5/2) and was characteristic of the high-spin state, provided that the zero field splitting parameters D and E are very small[13,14] The signal is illustrated in Fig. 5.

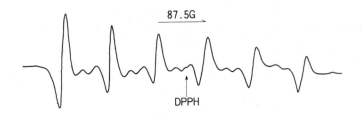

87.5G

DPPH

Fig. 5. ESR spectrum of Mn(II)-TPP in chloroform-imidazole solution at 77°K.

No change was observed in the spectrum of Fig. 5 on exposing the solution to oxygen at room temperature, indicating that the imidazole Mn(II) complex was not oxidized. Further observations were made on Mn(II)-TPP coordinated with other bases using either magnetic or optical technique. The results showed that the oxidative behavior of Mn(II)-TPP complexes was dependant on the axial ligand. Some representative results are shown below. It is assumed that dipyridine or diimidazole Mn(II) complex was formed in chloroform[15] although these complexes were not isolable.

Mn(II)-TPP (high)   ⟶ Mn(III)-TPP (high)[15]
Mn(II)-TPP-2Py (high) ⟶×⟶ Mn(III)-TPP-2Py (high)
Mn(II)-TPP-2Im (high) ⟶×⟶ Mn(III)-TPP-2Im (high)

Interaction of Oxygen with Cobalt(II) Complexes

All of the Co(II)-TPP complexes coordinated with axial ligands gave rise to ESR spectra characterized by eight hf lines typical of the cobalt($I = 7/2$). Some of them yielded three shf lines due to ligand nitrogen ($I = 1$). The presence of the three shf lines suggests that the cobalt complex is coordinated with a single ligand molecule on its z-axis, unlike Fe(II)-TPP or Mn(II)-TPP.

$$Co(II)\text{-}TPP + nPy \longrightarrow Co(II)\text{-}TPP\text{-}Py + (n-1)Py$$

A crystalline sample was isolated from the chloroform-pyridine solution on evaporation, and its composition was consistent with a monopyridine complex, or $C_{49}H_{33}N_5Co$; calcd., C : 78.38, H : 4.44, N : 9.32; found, C : 78.38, H : 4.22, N : 9.63. This pentaccordinated form would probably be due to the d^7 low-spin electronic configuration of Co(II).[3] The ESR spectrum for Co(II)-TPP coordinated with pyridine is shown in Fig. 6a.

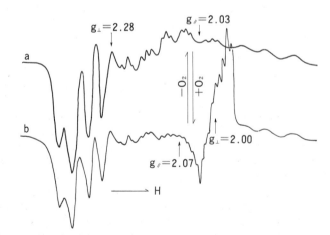

Fig. 6.   ESR spectra of Co(II)-TPP-Py in chloroform at 77°K. (a) evacuated; (b) exposed to O_2 for 10 min at room temperature.

When the Co(II)-TPP-Py complex was exposed to oxygen, the original signal due to Co(II) more or less decayed while a new signal developed in the vicinity of $g \sim 2$ (Fig. 6b). The spectral change was reversible with respect to varying oxygen pressures. The new signal near $g \sim 2$ showed eight shf lines, suggesting that a molecular oxygen ion is formed and combined with cobalt.

Oxygenation took place irreversibly with the Co(II)-TPP coordinated with imidazole. In this case no ESR signal was detectable after the irreversible oxygenation. The 3 d_{z^2} orbital energy of Co(II) would rise markedly when a strong base such as imidazole is coordinated. As a consequence, the unpaired electron of Co(II) may readily be transferred to the oxygenated cobalt complex, resulting in a diamagnetic, binuclear cobalt complex. The complexing of oxygen would then be no longer reversible.

SUMMARY

So far we have dealt with the oxidation of tetra-, penta- and hexacoordinated transition metal(II) complexes differing in spin state. In order to bring about a rapid oxidation, (1) the complexes should provide empty sites for oxygen, and (2) activation energy should be low. These conditions seem to be fulfilled, at least in the oxidation of Mn(II)-TPP, Fe(II)-TPP and Co(II)-TPP-Py. Conversely, the oxidation should be slow or inhibited if empty sites are not available and a significant energy is required for electron transfer. Such a case has been realized in the oxidation of Fe(II)-TPP-2Py. These examples are obviously the two extreme cases.

We have found an intermediary situation in the oxidation of the Fe(II)-TPP-2Im complex, where empty sites are no longer available, but the activation energy must be low enough because of the rapid oxidation that took place. We have presented a long range interaction mechanism between oxygen and Fe(II) ion through a conjugated porphyrin ring. Kinetic data for the oxidation was only qualitative but the interpretation was consistent with the electronic and structural properties of the complex.

Electron transfer via the so-called outer-sphere activated complex has been recognized in electron exchange reactions for

$$[Fe(CN)_6]^{4-} \rightleftharpoons [Fe(CN)_6]^{3-}$$
$$[Fe(phenanthroline)_3]^{2+} \quad [Fe(phenanthroline)_3]^{3+}$$

and so on.[6] A net chemical reaction is not involved here. The electron transfer reaction demonstrated above between oxygen and Fe(II)-TPP-2Im would thus constitute a new aspect in the problem of electron transfer.

References

1) K. Yamanoto, K. Marumo, K. M. Sancier, and T. Kwan, in Recent Developments of Magnetic Resonance in Biological System (S. Fujiwara and L. H. Piette, eds.), Hirokawa Pub., Tokyo, 1968, p.164.
2) Y. Ogata, K. Marumo, and T. Kwan, Chem. Pharm. Bull., 17, 1194 (1969).
3) K. Yamamoto and T. Kwan, J. Catalysis, 18, 354 (1970).
4) P. Rothemund and A. R. Menotti, J. Am. Chem. Soc., 70, 1808 (1948).
5) J. S. Griffith, in Molecular Biophysics (B. Pullman and M. Weissbluth, eds.), Academic Press, New York, 1965, p.191.
6) J. S. Griffith, Nature, 180, 30 (1957).
7) H. Kobayashi, M. Shimizu, and I. Fujita, This Bulletin, 43, 2335, 2342 (1970).
8) L. M. Epstein, D. K. Straub, and C. Maricondi, Inorg. Chem., 6, 1720 (1967).
9) R. Countryman, D. M. Collins, and J. L. Hoard, J. Am. Chem. Soc., 91, 5166 (1969).
10) T. H. Moss, A. J. Bearden, and W. S. Caughey, J. Chem. Phys., 51, 2624 (1969).
11) H. A. O. Hill, A. J. Macfarlane, and R. J. P. Williams, J. Chem. Soc., (A) 1704 (1969).
12) J. Fajer, D. C. Borg, A. Forman, D. Dolphin, and R. H. Felton, J. Am. Chem. Soc., 92, 3451 (1970).
13) M. Kotani, Rev. Mod. Phys., 35, 717 (1963).
14) R. D. Dowsing and J. F. Gibson, J. Chem. Phys., 50, 294 (1969).
15) L. J. Boucher, J. Am. Chem. Soc., 92, 2725 (1970).
16) W. L. Reynolds and R. W. Lumry, Mechanism of Electron Transfer, The Ronald Press, New York, 1966, p.13.

Description of the Interaction Effects of Chemisorbed Particles Based on the Surface Electronic Gas Model

M. I. TEMKIN

The Karpov Physical-Chemical Institute, Moscow

Many authors ascertained for different cases of chemisorption that one of the following equations was fulfilled in the region of medium coverages, depending on whether the adsorption equilibrium, the rate of adsorption, or the rate of desorption was studied. The equations are the logarithmic adsorption isotherm:

$$\theta = \frac{1}{f} \ln a_0 P \tag{1}$$

where θ is the degree of covering, p is the equilibrium pressure of the gas, f and a_0 are constants, the Zeldovich-Roginskii equation for the rate of adsorption:

$$r_a = k_a P e^{-g\theta} \tag{2}$$

where P is the pressure of the gas, k_a and g are constants, and the Becker - Langmuir equation for the rate of desorption:

$$r_d = k_d e^{h\theta} \tag{3}$$

where k_d and h are constants. Equations (1), (2), and (3) were considered conjointly for the first time in connection with the discussion of ammonia synthesis kinetics,[1] and the notation $\alpha = g/f$ and $\beta = h/f$ was then introduced, which enables one to rewrite the equations as follows;

$$r_a = k_a P e^{-\alpha f \theta} \tag{2a}$$

$$r_d = k_d e^{\beta f \theta} \tag{3a}$$

The analogy between Eqs. (2a) and (3a) is evident; Eq. (1) may be considered as a consequence of these as it results

from the condition for the adsorption equilibrium $r_a = r_d$.
It is seen then that

$$\alpha + \beta = 1 \tag{4}$$

If in any system one of Eqs. (1), (2a) or (3a) is obeyed,
it is natural to expect that the others are obeyed too. In
some systems the applicability of the whole set of equations
was verified experimentally, for instance with H on Pt
(electrochemical studies by Dolin and Ershler), with N_2 on Fe
and (adsorption measurements by Emmett).
Equation (1) together with

$$(\partial \ln P/\partial T)_\theta = \varepsilon/k T^2 \tag{5}$$

Equation (2) together with

$$(\partial \ln r_a/\partial T) = E_a/k T^2 \tag{6}$$

and Eq. (3) together with

$$(\partial \ln r_d/\partial T)_\theta = E_d/k T^2 \tag{7}$$

where ε is the heat of adsorption, E_a and E_d are activation
energies of adsorption and desorption, all the values being
per particle, k is the Boltzman constant, T is the absolute
temperature. These equations yield linear relationships

$$\varepsilon = \varepsilon^0 - C\theta , \tag{8}$$

$$E_a = E_a^0 + \alpha C\theta \tag{9}$$

$$E_d = E_d^0 - \beta C\theta \tag{10}$$

where

$$C = -kT^2 \frac{d f}{d T} \tag{11}$$

Since $df/dT < 0$, it follows from (11) that $C > 0$. The
values ε^0, E_a^0, and E_d^0 are the values ε, E_a and E_d at $\theta = 0$.
Relationships of the form of (8), (9), and (10) were
repeatedly found experimentally.
The decrease in the heat of chemisorption with coverage
can be quite large. For instance, calorimetric measurements

made in our laboratory by Ostrovskii[2] showed a decrease in
the heat of adsorption of oxygen on silver from 125 kcal/mole
on free surface to 80 kcal/mole on a surface containing one
O atom per two Ag atoms (larger amounts of oxygen are
adsorbed in a different form). When the change in the heat of
adsorption with coverage is so sharp, the heat motion cannot
markedly affect the relationship between ε and θ ; it must
differ but little from that which holds at T=O. Consequently
Eq. (1) cannot be explained on the same line taken in the
derivation of the term due to intermolecular forces in the
Van der Waals equation of state.

Boreskov and Khasin[3] studied the rate of isotopic exchange between oxygen adsorbed on silver and oxygen gas.
They found that the probability of exchange per unit time was
same for all sites of the surface; this result showed that the
surface was homogeneous. However, these data refer to the
from of adsorbed oxygen which exists at $[O]/[Ag] > \frac{1}{2}$,
where $[O]$ is the number of adsorbed oxygen atoms, and $[Ag]$
is the number of surface silver atoms, whereas calorimetric
data cited above refer to the form which exists at $[O]/[Ag] < \frac{1}{2}$
and which cannot be investigated with the method used by
Boreskov and Khasin, since the equilibrium pressure of O_2
over this form is negligibly small. A comparision of the
rates of decrease in the adsorption heat and increase in
the electronic work function for the region $[O]/[Ag] < \frac{1}{2}$
leads to the conclusion[4] that the decrease in the adsorption
heat is due mostly, to the interaction of O atoms on the
surface. Consequently, treatment of the relationships
discussed based on the concept of a heterogenous surface
cannot be applied to the system O on Ag. An analogous
conclusion is true for the system H on Pt. The electrochemical charging curve for smooth platinum in acid solution,
obtained by Ershler, does not differ materially from the
corresponding curve for platinum black, obtained earlier by
Frumkin and Shlygin. This result is difficult to explain if
the form of the charging curve, which corresponds to Eq.
(1), is due to heterogeneity of the surface. Boreskov and
Vasilevich[5] studying the isotopic exchange of hydrogen on
platinum found for most of the surface a heterogeneity which
was substantially less than would follow from the charging
curves.

On the other hand, in the case of industrial ammonia
synthesis catalyst, i.e., iron with K_2O and Al_2O_3 as pro-
moters, a study of the rate of exchange of nitrogen isotopes

at equlibirium of chemisorbed nitrogen with gaseous hydrogen
and ammonia, which was carried out in our laboratory by
Boreskova and Kuchaev, showed a large heterogeneity and
no interaction for the nitrogen chemisorbed on the iron
surface.

It is natural to suppose that in other cases also clean
metal surfaces are homogeneous and the heterogeneity of
surfaces of industrial catalysts is connected with the presence
of promotors or impurities. Then the model of a heterogeneous surface, being always formally applicable when
Eqs. (1) - (3) are obeyed, nevertheless does not correspond
to the physical nature of the phenomena under discussion for
clean surface.

In a recent publication[4] a simple physical model ac-
counting for the interaction of chemisorbed particles, suggested earlier[6], was somewhat generalized and used for
a discussion of properties of oxygen films on silver.

The model is based on the supposition that chemisorbed
particles feed their electrons into the surface layer of the
solid, or take electrons from it, forming at the surface a
kind of two-dimensional electronic gas or hole gas.

Each two electrons which go to the surface layer must
acquire a larger kinetic energy than the preceding ones to
satisfy the Pauli exclusion principle. If electrons are
extracted from the surface each two electrons must be taken
from a lower energy level than the preceding ones; the heat
of adsorption gradually decreases in both cases. A simple
calculation in the spirit of the Sommerfeld's theory of
metals, but for the two - dimensional case, leads to the
following relation[4]

$$\varepsilon = \varepsilon^0 - (\eta^2 h^2 L / 4 \pi m^*) \theta \qquad (12)$$

Here η is the effective charge acquired by an adsorbed particle at adsorption (elementary electric charge being taken as
unity), h is the Plank constant, m^* is the effective mass of
the electron in the solid (m^* is substituted for the mass of
electron m to account for the interaction of electrons with
the lattice), and L is the number of adsorption sites on the
unit surface.

The values $\eta = 1$ or $\eta = - 1$ to a transition of one electron
from adsorbed particle or to it, correspond respectively.
In both cases, according to Eq. (12), ε decreases linearly
with the increase in θ ; this is in agreement with Eq. (8).

Fractional values of the effective charge are supposed to be possible; they emerge when there is a superposition of two states, one of which includes a transition of electron. Such an effective charge must, generally speaking, depend on coverage, the change in it will be the less the larger the resonance energy of the superposing states. In the deriva-tion of Eq. (12) the effective charge was supposed to be constant; this simplifying assumption is justified by compari-son of Eqs. (12) and (8). According to them

$$C = \eta^2 h^2 L / 4 \pi m^*$$ (13)

In the surface electronic gas model the energy of the system is fixed by the number of adsorbed particles and does not depend on their arrangement. Thus the configura-tional entropy of the adsorbed layer coincides with that for the simple Langmuir adsorption and Eq. (12) leads to the adsorption isotherm

$$\ln ap = \eta^2 \frac{h^2 L}{4 \pi m^* k T} \theta + \ln \frac{\theta}{1 - \theta}$$ (14)

where a is a constant. For the region of medium coverages Eq. (14) approximates to the logarithmic isotherm (1). It is usually claimed that the term $\ln (\theta/(1-\theta))$ for $\theta \cong \frac{1}{2}$ may be neglected since it vanishes at $\theta = \frac{1}{2}$ (compare for instance 7). More correct procedure is to substitute for $\ln(\theta/(1-\theta))$ the first nonvanishing member of Taylor series

$$\ln \frac{\theta}{1 - \theta} = 4 (\theta - \frac{1}{2})+\frac{16}{3}(\theta - \frac{1}{2})^3+ \dots \frac{4}{2^{n+1}}(\theta - \frac{1}{2})^{2n+1}+ \dots$$ (15)

This gives Eq. (1) with

$$f = \eta^2 \frac{h^2 L}{4 \pi m^* k T} + 4$$ (16)

and

$$a_0 = ae^2$$ (17)

Values of f calculated with the help of Eq. (16) are of correct order of magnitude. Thus for platinum at 300°K if $\eta = +1$ the value f = 15 is obtained, using $m^*/m = 13$ from experimental data on electronic heat capacity[8] and L= 1.5 × 10^{15} cm^{-2}, which is the number of Pt atoms on 1 cm^2 of the face (III) i.e., the face of greatest density. According to electro-

chemical data, when H atoms are adsorbed on Pt surface
f = 14. Close values were obtained for adsorption of different
organic substances on platinum[9,10] For adsorption of ions I^-
and Br^- on platinum f = 13.8 and f = 15.5 respectively, were
obtained[11] The cited values of f are compatible with the
supposition that hydrogen is adsorbed on platinum in form of
H^+, and when ions I^- and Br^- are adsorbed, electrons pass to
platinum with the formation of covalent bonds Pt - I, Pt - Br.
      The simplest way to obtain the adsorption isotherm for
the case when a molecule X_2 dissociates in the act of adsorption into atoms X, is to substitute in Eqs. (1) or (13) for
p the pressure of X atoms corresponding to the equlibrium

$$X_2 = 2X \tag{18}$$

i.e., $P_X = (Kp_{X_2})^{1/2}$ where K is the equilibrium constant. We
obtain again an equation of the form of Eq. (1), but with f
doubled. Thus, at dissosiative adsorption

$$f = 2\eta^2 \frac{h^2 L}{4\pi m^* kT} + 8 \tag{19}$$

where η is the effective charge of an adsorbed atom X_{ads} .
The constant C in this case is also doubled as compares with
the value predicted by Eq. (13):

$$C = 2\eta^2 \frac{h^2 L}{4\pi m^*} \tag{20}$$

      For silver $m^*/m \cong 1$, $L = 0.7 \times 10^{15}\,cm^{-2}$ (one 0 atom per
two surface Ag atoms) and Eq. (20) gives $C = \eta^2 \times 78$ kcal/mole.
From experimental data cited above C = 45 kcal/mole, so
$\eta = \pm 0.75$. Taking into account the large electronegativity
of oxygen it is natural to choose the minus sign[4]
      To describe the velocity of a surface process some ef-
fective charge $\eta^{\neq}$ must be ascribed to the activated complex.
Taking it as independent of the covering one obtains for
adsorption and desorption of diatomic molecules of the type

$$X_2 = 2X_{ads}$$

Eqs. (9) and (10) with

$$\alpha = \eta^{\neq}/2\eta \tag{21}$$

·and with the corresponding $\beta = 1 - \alpha$.

We note that 2η is the effective charge of the two particles X_{ads}which are formed from X_2. Thus according to Eq. (21) α is the fraction of charge transfer which is accomplished when the activated complex is formed.

Kinetics of elementary reactions on catalytic surfaces can be considered similarly.

If two adsorbed molecules take part in a reaction, a generalisation of equation (18) for the case of simultaneous adsorption of two substances will be needed. The derivation of the corresponding equations can be achieved on the base of the physical notions discussed above.

References

1) M. I. Temkin and V. Pyzhev, Zhurn. fiz. khim (Russ.), 13, 851 (1939).

2) V. E. Ostrovskii and N. V. Kul'kova, Doklady AN SSSR (Russ.), 161, 1375 (1965).

3) G. K. Boreskov, A. V. Khasin, J. Res. Inst. Catalysis, Hokkaido Univ., 16, 447 (1968).

4) L. I. Shakhovskaya, L. A. Rudnitskii, N. V. Kul'kova and M. I. Temkin, Kinetika i Kataliz (Russ.), 11, 467 (1970).

5) G. K. Boreskov and A. A. Vasilevich, Kinetika i Kataliz (Russ.), 1, 69 (1960).

6) M.I. Temkin, Voprosy Khimicheskoi kinetiki, kataliza i reaktsionnoi sposobnosti (Russ.) Izd. AN SSSR, 1955, p. 484.

7) E. K. Rideal, Concepts in Catalysis. Academic Press, London - New York, 1968.

8) Ch. Kittel, Elementary Solid State Physics. John Wiley & sons, New York - London, 1962.

9) V. S. Bagotskii and Yu. B. Vasil'ev, Uspekhi elektro-khimii organicheskikh soedinenii (Russ.). "Nauka" 1966, p. 38.

10) S. Trassatti and L. Formaro, J. Electroanalitical Chem., 17, 343 (1968).

11) J. N. Pirtskhalava. Thesis. Moscow, 1970.

Catalysis by Electron Donor-Acceptor Complexes: Mechanism of Hydrogen Exchange Reactions on Electron Donor-Acceptor Complexes of Aromatic Compounds with Alkali Metals

K. Tamaru

University of Tokyo, Tokyo

When electron-donating and -accepting molecules are brought into contact, they generally form electron donor-acceptor (EDA) complexes, in which are combined charge-transfer interaction, electrostatic interaction, dipole-dipole interaction, etc. To various extents, an electron, or electrons, may be transferred from the electron donor to the acceptor. In an extreme case, complete electron transfer results in the formation of ion radicals or ion salts. In these EDA complexes specific dynamic properties, which are markedly uncharacteristic of the component molecules (such as light absorption), are of great interest.

When evaporated films of transition metal phthalocyanines were exposed to alkali metal vapor to form EDA complexes, it was discovered that not only the chemisorption of nitrogen, hydrogen and carbon monoxide, but also of ammonia, as well as Fischer-Tropsch syntheses proceeded.[1]

As has been demonstrated in previous papers[1,2] hydrogen is dissociatively adsorbed and some catalytic reactions, e.q., the hydrogen exchange reaction of molecular hydrogen (or acetylene), the selective hydrogenation of unsaturated hydrocarbons, and the isomerization of butenes, take place readily at room temperature over the electron donor-acceptor (EDA) complexes of some aromatic compounds with strong donors, such as alkali metals, metallocenes and phenothiazine, although no such reactions proceed over each of the components of the complexes.

In this report the mechanism of the hydrogen exchange reaction was studied in more detail over stoichiometric EDA complexes of aromatic compounds with alkali metals. The role of the chemisorbed hydrogen on the EDA complexes was examined and the active intermediates for the hydrogen exchange reactions were determined by various spectroscopic techniques.

EXPERIMENTAL

Stoichiometric films of anthracene complex with sodium ($An^{2-}2Na^+$ and An^-Na^+ : An=anthracene) were prepared from stoichiometric tetrahydrofuran (THF) solutions by evaporating the solvent, as described in previous papers.

The formation and composition of the complexes as films or in solution[3] were confirmed by their characteristic absorption spectra. The complex films were evacuated at about 100°C for one day prior to the experiments. Anthracene was purified by repeated recrystallization and by the zone-melting method, and the sodium was purified by distillation. In the hydrogen exchange reactions the isotopic hydrogen mixtures were analyzed by gas chromatography using an activated alumina column.

The reactions were carried out in a closed circulating system (300 cm³), which included a U-type reaction vessel.

RESULTS AND DISCUSSION

D_2- HZ* Exchange Reaction over 1:2 Anthracene-sodium Complex Film

When D_2 (20 cm Hg) was introduced onto the film of $An^{2-}2Na^+$, 2×10^{-3} mole at 80°C, a considerable amount of D_2 was adsorbed, and at the same time the gaseous deuterium was diluted by hydrogen from the complex forming HD in the gas phase, as shown in Fig. 1. After several days over such an EDA complex**; washing with D₂, the amount of HD produced and the amount of deuterium adsorbed became virtually constant, the final values suggesting that the reactions were not restricted to the surface of the complex film.

$$(D_2)_{ads} / (An^{2-}2Na^+)_0 = 0.9 \text{ for } 100 \text{ hr,}$$
$$(N_{exch}) / (An^{2-}2Na^+)_0 = 1.4 \text{ for } 100 \text{ hr at } 90°C$$

where N_{exch} and $(An^{2-}2Na^+)_0$ denote the number of hydrogen atoms exchanged with D₂ at saturation and the initial amount

---

* D_2-HZ denotes a type of hydrogen exchange reaction such as that between D_2 and the hydrogen of the anthracenesodium complex molecule.

** The $An^{2-}2Na^+$ was deposited on silica wool (the surface area was estimated to be ca. 40 m²).

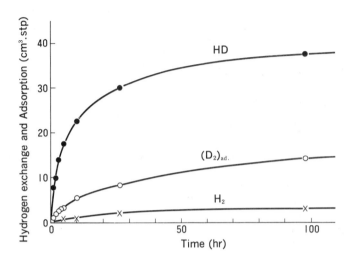

Fig. 1.   The kinetics of the D_2 - HZ exchange reaction and hydrogen adsorption (D_2) ads over An^{2-} $2Na^+$ complex film at 80°C, P_{D_2} = 40 cmHg.

of $An=2Na^+$, respectively.

The hydrogen adsorbed over the $An^{2-}2Na^+$ film could be completely removed by evacuating the system at 100°C for 20 hours, and reproducible results were obtained for the D_2-HZ reaction and hydrogen adsorption.

The pressure dependence of HD formation in the D_2-HZ exchange reaction was also studied over the complex film (evacuated before each run) at 85°C. It was found that the rate of HD formation on fresh film was less than first-order (ca 0.7) with respect to the D_2 pressure.

The behavior of hydrogen adsorbed on the $An^{2-}2Na^+$ film in the hydrogen exchange reaction was examined at temperatures between 25° and 85°C, by admitting D_2 (20 cmHg) onto a film (5×10^{-3} mole) on which various amounts of H_2 (from 2 to 42 cm³ stp) had been preadsorbed. The desorption of hydrogen was very slow at the lower temperatures. It was found from the results, as shown in Fig. 2, that at lower temperatures (25 to 70°C) the rate of HD formation decreased slightly with increasing amounts of preadsorbed hydrogen. On the

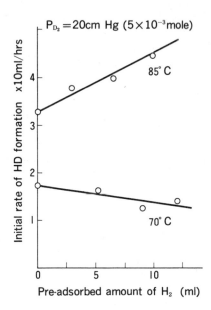

Fig. 2.   Dependence of the initial rate of H_2-D_2 exchange reaction upon the amount of pre-adsorbed H_2 over An^{2-} $2Na^+$.

other hand, at higher temperatures such as 85°C the rate increased slightly with small amounts of the preadsorbed hydrogen. In addition, the initial rates of HD formation for all temperatures at zero H_2 adsorption correspond to those observed over the fresh An^{2-} $2Na^+$ films. These results suggest that, at least in the initial stage, the D_2-HZ exchange reaction takes place directly between the gas phase D_2 and the hydrogen of the An^{2-} $2Na^+$ complex. This is also supported by the fact that, as shown in Fig. 1, the rate of HD formation was faster than that of D_2 adsorption.

The exchange reaction between D_2 (20 cm) and an almost saturated adsorbed hydrogen layer, ($(H_2)_{ads}$ /(An^{2-} $2Na^+)_0$ = 0.9) was studied at temperatures between 20° and 90°C. The rates of HD formation in the D_2-$(H_2)_{ads}$ exchange reaction and the activation energy (9.8 kcal·mole) were measured. The

rates of the D_2-$(H_2)_{ads}$ exchange reaction were linearly dependent upon the D_2 pressure (6.5 to 36 cmHg) at 62°C.

To determine the position of the hydrogen reacting in the
D_2-HZ exchange, the complex film was decomposed by oxygen after exchange with D_2, and the resultant anthracene was
analyzed by mass spectrometry. The results suggested that
a mono-deuterated (9-position of the anthracene) product was
obtained in the initial exchange, but di-deuterated anthracene
(9- and 10-positions) was the major product after complete
exchange with D_2.

As was indicated previously,[2] 9-monohydroanthracenium
sodium (AnH Na$^+$) and sodium hydride were produced by
hydrogen adsorption over an $An^{2-}2Na^+$ film.

$$\text{OOO}^= 2Na^+ + H_2 \longrightarrow \text{O}\overset{H\,H}{\underset{\times}{U}}\text{O}^- Na^+ \cdot NaH$$

AnH$^-$Na$^+$ was also independently prepared by reacting
$An^{2-}2Na^+$ and 9,10-dihydroanthracene (AnH$_2$) in THF at 25°C,[2]

$$An^{2-}2Na^+ + AnH_2 = 2AnH^-Na^+$$

The AnH$^-$Na$^+$ thus prepared, (identified by its characteristic
electronic absorption band at 432 mμ) was deposited on the
walls of the reaction vessel (270 ml) by evaporating the solvent at 25° C. The film of AnH$^-$Na$^+$ was stable at temperatures below 80°C. When 20 cmHg of D_2 was introduced onto
the AnH$^-$Na$^+$ film at temperatures between 20° and 80°C, the
reversible HD formation obeyed first-order kinetics, and
attained an equilibrium value. It was also found that at 65°C
the rate of HD formation in the D_2-AnH$^-$Na$^+$ exchange reaction
was proportional to the pressure of D_2 (6.5 to 45 cmHg), and
to the amount of AnH$^-$Na$^+$. In addition, it was suggested that
at temperatures between 25° and 65°C and at a D_2 pressure of
20 cm, an AnH$^-$Na$^+$ film (3.8×10^{-3} mole) gave a D_2-AnH$^-$Na$^+$
exchange rate (cm^3/hr·mole) similar to that obtained for the
D_2-$(H_2)_{ads}$ exchange reaction over an equal weight of $An^{2-}2Na^+$
which was almost saturated with adsorbed H_2.

AnH$^-$Na$^+$ dissolved in d_8-tetrahydrofuran(98% purity,
Merck Co.,), was completely exchanged with D_2 at 65°C, and
the complex solution analyzed by nmr spectrometry. It was
found that deuterium exchange took place selectively at the
10-position of the 9-monohydroanthracenium sodium, and a
small amount of monodeuterated 9,10-dihydroanthracene was
formed in the resultant solution. The interpretation of the

spectra was carried out using the results obtained by
Schneider.[5]   The deuterium exchanged monohydroanthracene
was evacuated (AnH^-Na^+ = An^-Na^+ + $1/2$ H_2), decomposed by
oxygen, collected by sublimation, and analyzed by mass
spectrometry. This demonstrated that monodeutero (9-po-
sition) anthracene was the major product.

$$d_0 = 10 \%, \quad d_1 = 88 \%, \quad d_2 = 2 \%, \quad \text{and} \quad d_3 = 0 \%$$

These results seem to indicate that the monohydro-
anthracenium complex participates in the hydrogen exchange
reaction at the carbon atom opposite the one where hydrogen was added, and during the course of the hydrogen exchange reaction desorption of the added hydrogen takes place
rather slowly. Thus, only the d_1 species is produced to any
extent by deuteration, and no marked mixing at two equivalent
carbon atoms (9- and 10-positions) was observed. From
these results it was suggested that the D_2-AnH^-Na^+ exchange
reaction proceeds as follows;

To clarify the role of NaH in the hydrogen exchange
reaction, a little excess sodium hydride (75 mg, 3×10^{-3}
mole) was added to AnH^-Na^+ (2.8×10^{-3} mole) in THF, and a
mixed film of AnH^-Na^+ and NaH^* was thus deposited on silice
wool. The initial rates of HD formation in the temperature
range 20°to 68°C were not appreciably different from those
obtained over the AnH^-Na^+ film alone (2.8×10^{-3} mole).

It is thus suggested that the presence of NaH does not
essentially affect the rate of the hydrogen exchange of AnH^-Na^+
with D_2; the small increase in the rate of HD formation
observed on adding excess NaH seems to be due to an increase in the surface area of the AnH^-Na^+ when it is deposited
on the NaH powder.

To confirm this conclusion, the hydrogen exchange
reaction was carried out in three different systems as
follows,

* The hydrogen exchange reaction did not proceed over the
NaH alone under these reaction conditions.

$$(AnH^-Na^+, NaH) + D_2 \xrightarrow{r_1} HD$$
$$(AnH^-Na^+, NaD) + D_2 \xrightarrow{r_2} HD$$
$$(AnH^-Na^+, NaD) + H_2 \xrightarrow{r_3} HD$$

where AnH^-Na^+ (3×10^{-3} mole) and NaH or NaD* (ca. 1 g, 4×10^{-2} mole) were employed to make the mixed films.

The rates r_1 and r_2 were found to be almost equal (ca. 4 cm^3/hr), whereas r_3 was negligibly small at 65°C under the same reaction conditions.

When an adduct film of AnH^-Na^+ (3×10^{-3} mole) and NaD (ca. 4×10^{-2} mole) was heated at 85°C for several hours in vacuo, a small amount of hydrogen gas was collected and a gas chromatographic analysis gave its composition as:

$$H_2 = 2\%, \quad HD = 87\% \text{ and } D_2 = 11\%$$

From UV-spectroscopic measurements it was found that $An^{2-}2Na^+$ was regenerated and AnH^-Na^+ decreased by eliminating hydrogen,

$$AnH^-Na^+ + NaD \longrightarrow An^{2-}2Na^+ + HD$$

These results suggest that it is the hydrogen in the AnH^-Na^+ which participates in the exchange with molecular hydrogen, the hydrogen of NaH being unreactive for the exchange reaction at the temperatures studied.

It was therefore concluded that the overall process for the D_2-H_2 exchange reaction occurred simultaneously with hydrogen adsorption on the $An^{2-}2Na^+$ film, as follows;
i) Direct hydrogen exchange between $An^{2-}2Na^+$ and D_2 accompained by dissociative chemisorption;[**]

* 98% deuterated sodium hydride was obtained by repeated exchange of NaH with D_2 for several days at 210°C.
**The localization energy of the π electron at the carbon where hydrogen exchange takes place is similar in both i) and ii), -1.18 and -1.30 β respectively (calculated from the simple Huckel m. o.), whereas for neutral anthracene it is -2.01 β at the 9 or 10-position.

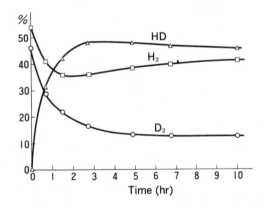

ii) Direct hydrogen exchange of D_2 with AnD^-Na^+, which was formed by hydrogen adsorption;

$$\text{HD} \atop \underset{D}{\boxed{O|U|O}} Na^+\cdot NaD + D_2 \longrightarrow \underset{D}{\overset{H\ D}{\boxed{O|U|O}}} Na^+NaD + HD$$

iii) At temperatures above 85°C the desorption process also contributes to the hydrogen exchange reaction;

$$\overset{H\ D}{\boxed{O|U|O}}^- Na^+NaD \longrightarrow \overset{D}{\boxed{O|O|O}}^{2-} 2Na^+ + HD$$

The H_2-D_2 Exchange Reaction over 1:2 Anthracene-sodium Complex Film.

When a 1:1 mixture of H_2 and D_2 ($P_{H_2+D_2}= 20$ cmHg) was admitted onto a fresh $An^{2-}2Na^+$ film (5.2×10^{-3} mole) deposited on silica wool, HD was produced as shown in Fig. 3.

Fig. 3.   The kinetics of the H_2-D_2 exchange reaction over an $An^{2-}2Na^+$ film (5.2×10^{-3} mole) at 82°C. $P_{H_2+D_2}= 20$ cmHg.

The initial rate of HD formation in the H_2-D_2 exchange reaction was compared with that of the D_2-HZ reaction and

the following conclusions were derived. At higher temperatures, such as 82°C, HD formation in the H_2-D_2 exchange
reaction resulted mainly from the following two exchange
reactions which involve the bonded hydrogen of the $An^{2-}2Na^+$
complex;

$$D_2 + HZ = HD + DZ$$
$$H_2 + DZ = HD + HZ \qquad HZ = An^{2-}2Na^+ \text{ or } AnH^-Na^+$$

At these higher temperatures the initial rates of HD formation from an H_2-D_2 mixture were similar to those of the
D_2-HZ exchange reaction (P_{D_2} = 20 cmHg) as shown in Fig. 4.

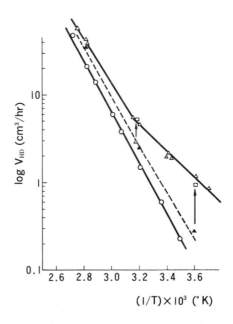

Fig. 4. Arrhenius' plot for HD formation in the H_2-D_2 exchange reaction over An^{2-}-2Na films ($\triangle$). and in the D_2 -
HZ (HZ = $An^{2-}2Na^+$) exchange reaction ($\bigcirc$). The rates
of HD formation in the H_2-D_2 exchange reaction
after the preadsorption of CO (Ca. 4.5 ml) ($\blacktriangle$), evacuation
of the preadsorbed CO from the complex film ($\square$).

On the other hand, at lower temperatures such as 42°C, HD formation from an H_2-D_2 mixture was faster than in the D_2-HZ exchange reaction; this is shown in Fig. 4.

The HD formation could not be explained by the above bonded mechanism alone, and the H_2-D_2 exchange reaction almost reached thermal equilibrium in the gas phase. The pressure dependence of the rate of the H_2-D_2 exchange reaction at 45°C was approximately proportional to the total pressure of the H_2-D_2 mixture (1:1), and to the square root of each of the H_2 and D_2 pressures.

$$V_{HD} \propto kP_{H_2}^{\frac{1}{2}} \cdot P_{D_2}^{\frac{1}{2}}$$

From these results at lower temperatures it was suggested that the H_2-D_2 exchange reaction took place mainly via the recombination of "mobile" hydrogen chemisorbed on the complex film:

$$D_2 \rightleftharpoons 2D\ (a)$$
$$H_2 \rightleftharpoons 2H\ (a) \qquad 2HD\ (a) \rightleftharpoons 2HD$$

It was also of interest to note that when the $An^{2-}2Na^+$ film adsorbed CO (4.5 and 16 cm^3 stp) at 40°C, HD formation from the D_2-HZ exchange reaction was not affected, but the rate of the H_2-D_2 exchange reaction decreased markedly, and the initial rates of HD formation became almost equal to those obtained at similar temperatures for the D_2-HZ exchange reaction. After evacuation of the complex film at 85-100°C for 20 hr the activity of the film for the H_2-D_2 exchange reaction was completely restored, as shown in Fig. 4. It was suggested that CO effectively inhibited the "mobile chemisorbed hydrogen" mechanism, but had negligible effect on exchange between D_2 and the hydrogen of the complex molecule.

Hydrogen exchange reaction in THF solution.

The EDA complexes of various aromatic compounds with alkali metals can also exchange their hydrogen with molecular hydrogen in THF solution. The results obtained over the complexes of anthracene with alkali metals are given in Table I. It was clearly demonstrated that the activity of those EDA complexes in solution depends markedly upon the alkali metals employed. Hydrogen absorption takes place in solutions of $An^{2-}2Li^+$ and $An^{2-}2Na^+$, forming monohydroanion and sodium hydride, whereas in the case of $An^{2-}2K^+$ and $An^{2-}2Rb^+$,

Table I

Complex (5 × 10⁻³ mole)	Solvent (80 ml)	Hydrogen absorption		Hydrogen exchange	
		$V^*_{D_2 ads}$ (ml/hr) 27°C	E (kcal/mole)	V^*_{HD} (ml/hr) 27°C	E (kcal/mole)
$An^{2-}2Li^+$ {	THF	6.8	10.8	0.05	——
	DME	6.3	11.0	0.03	——
$An^{2-}2Na^+$ {	THF	5.7**	11.0	0.12	10.5
	DME	4.2	11.4	0.08	——
$An^{2-}2K^+$	DME	less than 0.01	——	2.40	12.0
$An^{2-}2Rb^+$	DME	less than 0.01	——	1.35	12.8
AnH^-Li^+	THF	less than 0.01	——	0.07	——
AnH^-Na^+	THF	less than 0.01	——	0.08	9.8
AnH^-NaNaH^{***}	THF	0	——	0.08	9.8

a reversible hydrogen exchange proceeds without appreciable hydrogen absorption, as shown in Figs. 5 and 6.

In all cases the H_2-D_2 exchange reaction proceeds by the repetition of the following two steps:

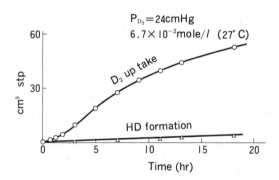

Fig. 5. Hydrogen adsorption and hydrogen exchange by $An^{2-}2Na^+$ in THF solution.

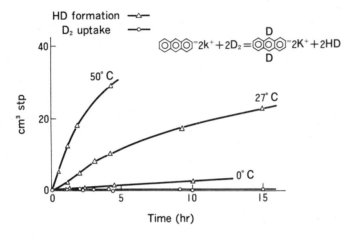

Fig. 6.   Hydrogen adsorption and hydrogen exchange reaction by An^{2-} $2K^+$ in DME solution.

$$D_2 + HZ = HD + DZ$$
$$\underline{H_2 + DZ = HD + HZ}$$
$$H_2 + D_2 = 2HD$$

where HZ represents the EDA complexes of $A^- Na^+$ and $A^{2-} 2Na^+$. This corresponds to the hydrogen exchange i) which is mentioned above.

In a similar manner the D_2-HZ exchange reaction over the EDA complexes of potassium with various aromatic compounds was studied in THF solution, and the rate of HD formation was correlated with the localization energies of their π-electron (or electrons), calculated by the simple Hückel m.o., as given in Fig. 7.   It can be seen that the lower the localization energies, the faster the rate of HD formation.

The H_2-D_2 exchange reaction on various transition metals has been extensively studied for many years, but, nevertheless, much more information is still required before the true mechanism can be elucidated.   On the other hand, the catalytic activity of EDA complexes for the exchange reaction can be discussed on a molecular basis and may be correlated with the electronic properties of the component aromatic compounds, presenting a well-defined model case of a cataly-

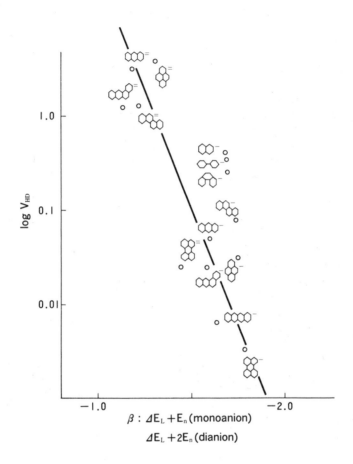

Fig. 7. Dependence of the rate of hydrogen exchange between the EDA complexes $(5 \times 10^{-3}$ mole) of potassium with various aromatic compounds upon the lowest localization energy of π electron (or electrons) of the aromatic anions. $P_{D_2} = 20$ cmHg. ΔE_L; localization energy of netural molecule; E_n : highest occupied level of neutral molecule.

tic exchange reaction.

References

1) K. Tamaru, Advan. Catalysis, 20, 327 (1969), Catalysis
   Review, 4, 161 (1970).
2) M. Ichikawa, M. Soma, T. Onishi, and K. Tamaru,
   Trans. Farad Soc., 63, 997, 1215, 2015 (1967).
   J. Catalysis, 6, 336 (1966), 9, 418 (1968), Bull. Chem.
   Soc., Japan, 40, 1015, 1294 (1967), 41, 1739 (1968).
3) M. Ichikawa, M. Soma, T. Onishi, and K. Tamaru, J.
   Am. Chem. Soc., 91, 6505 (1969).
4) M. Ichikawa, M. Soma, T. Onishi, and K. Tamaru,
   Bull. Chem. Soc., Japan, 43, 3672 (1970).
5) W. G. Schneider, in Nucl. Magnetic Resonance Chem.
   Proc. Symp., Cagliari, Italy, 1964.

Experimental Investigation of the Contribution of Charge Transfer from Complexes to Catalysts

O. V. KRYLOV

Institute of Chemical Physics, Academy of Sciences of the USSR, Moscow

The possible contribution of charge transfer complexes in heterogeneous catalysis has been discussed in a review.[1] It was shown, in particular, that the observed two-peak change of catalytic activity in the series of transition metal oxides may be due not only to changes in the crystal field stabilization energy as suggested by Dowden and Wells,[2] but also to a regular change in the energy of charge transfer (of an electron) from the ligand (adsorbed molecule, donor) to the transition metal atom. Variations in charge transfer energy in a series of similar complexes also show a two-peak dependence. It may be noted that such a dependence in the series of transition metal oxides is observed for redox reactions only.

Certain results on the structures of active sites at the surface of transition metal oxides and on those of complexes formed by transition metals and adsorbed molecules have been obtained lately at the Institute of Chemical Physics by the Catalysis Laboratory. Electron transfer was found to contribute both to adsorption and to catalysis.

The structures of active sites on molybdenum oxide, cobalt, vanadium and tungsten catalysts supported by BeO, MgO, and Al_2O_3 have been studied by the diffuse reflectance and ESR techniques. The above systems represent model oxidation catalysts.

It appeared[3] for MoO_3 supported by MgO and Al_2O_3 that the coordination of surface Mo^{6+} ions changes from tetrahedral to octahedral. This conclusion was made after observation of the charge transfer band $O^{2-} \longrightarrow Mo^{6+}$. With increase of the coordination number from 4 to 6 the absorption maximum shifts from 260 to 320 mμ.

Mo ions of lower valence were formed by reduction of the molybdenum oxide catalysts. Figure 1 shows optical spectra of reduced molybdenum oxide supported by MgO and Al_2O_3. The absorption maxima were identified and the

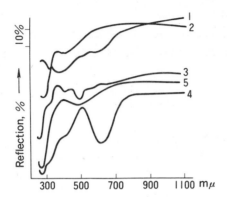

Fig. 1.   Optical diffuse reflectance spectra of reduced
molybdenum oxide (2%) supported by Al_2O_3 (1-2) and by
MgO (3-5). Samples 1 and 4 were obtained from sulfates
and contains sulphur admixtures.

structures formed were ascribed[4] to Mo^{6+} and Mo^{5+} in the
tetrahedral case to Mo^{5+} in the square pyramidal case and to
Mo^{4+}(or Mo^{3+}) in the octahedral case.

Investigation of the ESR spectra of Mo^{5+} in MgO and
Al_2O_3 revealed the presence of these ions in tetrahedal and
square-pyramidal coordination.[5]

The reduction chain can be written as

$$Mo^{6+}(T_d) \longrightarrow Mo^{5+}(T_d) \longrightarrow Mo^{5+}(S_p) \longrightarrow Mo^{5+}(O_h) \longrightarrow Mo^{4+}(O_h)$$

Evidently, the structures of $MgMoO_4$ and $Al_2(MoO_4)_3$ molybdates disintegrate in the course of reduction and solid solutions form, involving Mo ions of a lower degree of oxidation, O_h, such as Mo^{5+}, Mo^{4+}, and possibly Mo^{3+}, that may
substitute for Mg or Al ions in the support lattice. The
energy of charge transfer transition $O^{2-} \longrightarrow Mo^{6+}$ decreases
in the order Mg-0 -Mo, Al-0-Mo, Mo-0-Mo. For this reason
the stabilization of higher oxidation levels is easier with MgO
than with Al_2O_3 and molybdenum oxide proper. The reduction
of MoO_3 is accelerated by impurities introduced into the support.

It will be noted also that the ESR signal does not appear
at a Mo concentration lower than $10^{12}cm^{-2}$. This shows the

presence of paired, closely located Mo ions, for example Mo-0-Mo that reduce more readily. The Mo^{6+} ions might be reduced to Mo^{4+} with subsequent disproportionation

$$Mo^{6+} + Mo^{4+} \longrightarrow 2Mo^{5+}$$

The results obtained seem to reveal a possible contribution from two-electron (two-equivalent) transfers to oxidative conversion at the surface of molybdenum oxide catalysts.

The overall picture of stabilization of V ions on the surface and of the changes they overgo on reduction resembles that for Mo. The predominant part of V^{4+} ions on MgO and Al_2O_3 exist in the vanadyl configuration.

The lower V valences on MgO were represented by V^{2+} ions (g = 1.981, A = 75 gauss). It did not appear possible to reduce V-MgO and V-Al_2O_3 samples containing less than 5×10^{13} ions of V^{5+}/cm^2. It may be suggested by analogy with Mo catalysts that formation of polynuclear sites is necessary for reduction of V^{5+} and stabilization of V^{4+}. Probably, in addition, the 16 gauss splitting of lines A can be assumed to be due to the vanadyl ion coming into a paired active site of type $VO^{2+}...V^{5+}$.

Diffuse reflectance spectra of cobalt oxide supported by MgO and Al_2O_3 show that the greater part of Co on Al_2O_3 is present as Co^{2+} in a tetrahedral coordination, i.e. as a spinel $CoAl_2O_4$, and that on MgO as Co^{3+} in an octahedral coordination. Bands of two-valent octahedral Co are observed for MgO only at high concentrations of CoO. A gradual conversion of Co^{3+} to Co^{2+} is observed after heating of the CoO-MgO system to high temperatures of 1000°-1200°C. The Co^{3+} ions seem to stabilize on the surface layer at the expense of the OH group. Disappearance of OH groups from the surface favours the conversion from Co^{3+} to Co^{2+}.

The affinity of transition metal ions for various coordination numbers in a matrix of non-transition metals has been explained in terms of the crystal field stabilization energy[6].

The coordination and valency of ions in complex catalysts can differ from those in simple catalysts. It was found, for instance[7,8] that in an Al_2O_3 supported mixture of Co and Mo oxides, both ions are in a tetrahedral structure, as follows from data on differential interaction between CoO and Al_2O_3 ($\longrightarrow CoAl_2O_4$) and between MoO_3 and Al_2O_3 ($\longrightarrow Al_2(MoO_4)_3$). A complex catalyst accelerates dismutation of olefins and hydrosulphurization, whereas in a catalyst for partial propene

oxidation to acrylic acid, i.e. cobalt molybdate, both Co and
Mo ions are located mostly in an octahedron, i.e. they show
a completely different coordination. However, recent results
obtained by Lipsch and Schnit[9] show that at a CoMoO₄ surface
the Co ion will also be in terahedral coordination. Apparently, lower coordination numbers are a general characteristic
of surface ions and this ensures the possibility both of addition of new ligands and of the catalytic reaction.

The Co^{2+} ions in a low-spin electron configuration $t_{2g}^6 2g$
have one unpaired electron on the 2g orbital, i.e. they
display the properties of a free radical. With a Co-Mo-Al₂O₃
olefin dismutation catalyst, the Mo ion seems to repesent the
active site. The part played by the Co ion seems to be the
inhibition of surface condensation leading to coking of the
surface.

The diffuse reflectance spectra of various 2nd group
molybdates such as Be, Mg, Ca, Sr, Ba, Zn, Cd, show,
along with a small absorption maximum at 11,000 cm^{-1}, an
intensive absorption band (that seems to involve charge
transfer) at 20,000-23,000 cm^{-1} (Fig. 2). The energy cor-

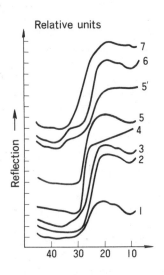

Fig. 2.　　Optical diffuse reflectance spectra of molybdates:
MgMoO₄ (1); BeMoO₄(2); ZnMoO₄(3); CdmoO₄ (4); BamoO₄(5);
heated at 550°C; BamoO₄ (heated at 750°C); SrMoO₄ (6);
CaMoO₄ (7).

responding to this band is considerably lower than for the
MoO_4^{2-} ions in aqueous solutions and for molybdates of alkali
metals. Seemingly, a common orbital (or zone) is formed
from the levels of the cations and molybdate ions, and
charge transfer from this orbital to the d-orbital, or from
the d-orbital to a relevant antibonding orbital becomes much
easier. Such a lowering of charge transfer energy might
also occur for the adsorption of various molecules on
molybdates. In this case electron transfer from the transition
metal ion to the adsorbed molecule over relatively long distances becomes possible.

Such electron transfer was observed experimentally by
the authors, together with Spiridonov and Pariiskii, in studying the radical form O_2^- of oxygen on the surface of supported
catalysts.[10]

Adsorption of O_2 on partially reduced molybdenum oxide
and vanadium oxide catalysts supported by MgO and Al_2O_3,
and also by BeO, yields O_2^- ion-radicals at room temperature.
However, they stabilize on the matrix Mg^{2+} and Al^{3+} atoms,
rather than on Mo and V atoms. Under certain conditions of
O_2 adsorption on Mo catalysts both the O_2^- and Mo^{5+} signals
become more intense. This indicates that the Mo^{4+} ion,
releasing an electron to the O_2 molecule through the support,
is responsible for O_2 adsorption as O_2^-. When O_2 is adsorbed
at $77°K$ O_2^- ion radicals stabilized on Mo^{6+} and Mo^{5+} ions, or in
their vicinity, are formed along with the above O_2^- ion radicals.
After the system is heated to room temperature O_2^- ion radicals
remain on the matrix ions only. Figure 3 shows the spectra
of O_2^- ion radicals adsorbed on Mo catalysts at room temperature. The data obtained were interpreted in terms of a relation between the anisotropy of the O_2^- g factor and the cation
charge,[11,12] and those for V catalysts also in terms of the HFS
of paramagnetic species adsorbed on surface V ions.
Moreover, signals similar to those of O_2^- on Mo-MgO and Mo-
Al_2O_3 catalysts at room temperature were obtained after
radiation chemisorption of O_2 on MgO and Al_2O_3 supports.
The scheme of electron transfer is

$$Mo^{4+}\text{-}O\text{-}Mg^{2+}\text{-} \xrightarrow[-196°C]{O_2} Mo^{5+}\text{-}O\text{-}Mg^{2+} \xrightarrow[-196°C]{\overset{O_2^-}{O}}$$

$$\overset{O_2^-}{Mo^{6+}}\text{-}O\text{-}\overset{O_2^-}{Mg^{2+}}\overset{e}{\text{-}}\overset{O_2^-}{Mg^{2+}} \longrightarrow \overset{O_2^-}{Mo^{6+}}\text{-}O\text{-}\overset{O_2^-}{Mg} \xrightarrow[25°C]{O} Mo^{5+}\text{-} O \text{-} \cdots$$

$$\overset{O_2^-}{}$$

$$\cdots O - Mg^{2+} \tag{I}$$

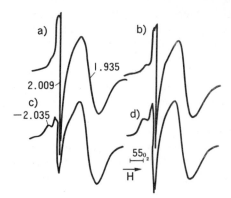

Fig. 3.   Low-temperature adsorption of oxygen on Mo/Al₂O₃.
(a) Initial ESR spectrum of O_2^- ion(oxygen adsorbed at 77°K;
(b) spectrum of the same sample after heating to room
temperature in vacuum; (c) sample heated in oxygen at 20°C,
with subsequent pumping off of oxygen; (d) sample heated in
O_2 at 100°C.

Electron transfer through the support lattice from the
transition metal ion to an oxygen molecule stabilized at
another active site of the catalyst was thus established. It
probably proceeds through Al-O-Mo, Mg-O-Mo, and
Mg-O-V bridges over the system of d-levels. Such an
electron transfer also seems to be possible from transition
metal ions inside the catalyst close to its surface, though on
the whole the catalyst is a dielectric.

Similar electron transfer with formation of O_2^- ion radicals
stabilized on Mg and Al ions of the matrix has also been
observed recently for cobalt oxide and tungsten oxide
catalysts supported by MgO and Al₂O₃.

Analogies to the results obtained can be found in
homogeneous catalysis. For instance, with homogeneous
liquid phase oxidation of olefins, palladium does not change
its oxidation state in the presence of Pd and Cu salts and is
only an intermediate agent for electron transfer.[15]

The O_2^- complexes formed directly with Mo and V ions are
very weak, probably due to symmetry hindrance. Data on
the surface structure of adsorbed olefin complexes have also
been obtained.

According to Fokina, reduction of supported molybdenum oxide catalysts results in a considerably higher amount of adsorbed olefin and in strengthening of its bonding. A strong π-complex is formed. When propylene is adsorbed on cobalt oxide catalysts[8], diffuse reflectance spectra show the same tetrahedral coordination of Co, with a certain change in intensity. In this case propylene is adsorbed on an oxygen ion bound to Co^{3+}, followed by propylene oxidation and conversion of Co^{3+} to Co^{2+}.

No bands from olefin complexes, but only those from products of olefin polymerization, were observed in most IR-spectra of olefins adsorbed on transition metal oxides[16].

Tabasaranskaya and Kadushin of the Institute of Chemical Physics observed no absorption bands in IR spectra from adsorbed olefins after adsorption of ethylene, propylene and butylene on NiO-MgO catalysts. Neither were such bands observed with consecutive adsorption of oxygen after the olefin or of the olefin after oxygen. Intensive infrared adsorption spectra were observed for simultaneous adsorption of the olefin and of oxygen. The intensity of absorption bands from acetylene, ethylene, propylene, butylene or butadiene mixtures with oxygen increases out of proportion to increasing concentration of Ni^{2+} in MgO, and this seems to be evidence of a possible contribution of binuclear sites to adsorption. With MgO proper, no surface compounds are seen in the IR-spectra.

The intense absorption bands in the 1600 cm^{-1} region were ascribed to a stretching vibration of the -C=C bond in the strongly adsorbed complex:

$$HC=CH \qquad HC=CH-CH_2 \quad \text{or} \quad H_2C=CH-CH_2$$

$$HC=CH-CH=CH$$

Less intense absorption bands in the 1,500 cm^{-1} region were ascribed to formation of π-complexes

$$H_2C=CH_2 \qquad\qquad H_2C=CH-CH_3$$

$$H_2C=CH-CH=CH_2 \qquad \text{and} \qquad H_2C=CH-CH-CH_2$$

Oxygen plays an important part in the oxidative abstraction of the hydrogen atom upon formation of strong surface
compounds of type III. Bands corresponding to carboxylates,
products of complete olefin oxidation, are also observed in
the IR-spectra. It will be of interest to note that with
butadiene adsorption (different from olefins with only one
-C=C bond) the IR spectrum is observed in the absence of
oxygen as well, the absorption bands being very close to the
spectrum obtained for adsorption of butylene with oxygen.
Butadiene might be the first intermediate in butylene oxidation.

Similar results were obtained for the CoO-MgO catalyst.
With MoO_3-MgO and V_2O_5-MgO catalysts the IR spectrum of
an adsorbed olefin-oxygen mixture appears only after the
catalyst is heated in a gas mixture at 200-300°C. When use
is made of an NiO-MgO catalyst with additional MoO_3 or
V_2O_5, the intensity of bands ascribed to π-complexes increases with the concentration of MoO_3 or V_2O_5, passes
through a maximum, and then drops.

The IR-spectrum of acetylene after its adsorption on the
NiO-MgO catalyst was ascribed to a surface compound

$$M-C=C-H$$

The formation of an allyl π-complex was often suggested
to be the first stage of partial olefin oxidation.

The presence of an unpaired electron on the d_{xz} or d_{yz} orbital
of the transition metal is necessary for the formation of such
complexes, and thus the elements at the beginning of the
transition series are particularly favourable. When propylene is adsorbed on bismuth molybdate, an increase and
then a decrease in the intensity of the Mo^{5+} signal is observed
in the ESR spectrum.[7] The authors suggest the formation of
an intermediate allyl π-complex

$$C_3H_6 + Mo^{6+} + O^{2-} \longrightarrow [C_3H_5 \cdot Mo^{6+}] + OH^- + e$$

the electron being transferred to the neighbouring Mo^{6+}ion, which is reduced to Mo^{5+}. Further allyl oxidation results in reduction of $Mo^{5+}$to Mo^{4+}. The existence of intermediate allyl complexes clearly follows from isotopic data[8] For example, all the products of catalytic oxidation of propylene which is labeled with ^{14}C: acrolein, acetaldehyde, formaldehyde, CO_2, show the same radioactivity. This is evidence that the carbon atoms in the intermediate complex are indistinguishable.

It will be noted, however, that virtually no direct physical measurements revealing the presence of π allyl complexes on the surface have been made. The absence of an ESR signal from the allyl radical was accounted for[17]by formation of a diamagnetic complex $C_3H_5...Mo^{6+}$, Mo being bound to the allyl by a σ-π-bond completely occupied with electrons.

A paper on IR-spectra obtained for adsorption of propylene on ZnO has been published recently.[18] The 1505 cm^{-1} absorption band was ascribed to the stretching vibration of the C=C bond in the π allyl complex, by analogy with homogeneous allyl complexes of manganese. However, it could be ascribed with greater probability to vibration of the C=C bond in a simple π-complex. Vibration in the π allyl complex would be shifted by 30-50 cm^{-1}towards the region of shorter wavelength. The above research of Tabasaranskaya and Kadushin, dealing with adsorption of olefin-oxygen mixtures on MgO, has shown no π allyl complexes.

Extensive research concerned with adsorption of hydrocarbon-oxygen mixtures on catalysts for partial oxidation[8] shows that under conditions close to those for catalysis the amount of adsorbed oxygen and propylene, and of their mixtures, is considerably higher than that observed for differential adsorption, and sometimes is higher than that for consecutive adsorption.

The formation of intermediate hydrocarbon-oxygen complexes was found by isotopic methods, by measurement of the electron work function, and by use of the Mössbauer effect.

The appearance of bands from type II complexes in the presence of oxygen may be ascribed to stronger polarization of olefin molecules in the presence of oxygen, or to electron transfer from the olefin to the adsorbed oxygen, similar to that of type I complexes of Mo or V with O_2 adsorbed on Mg or Al. The positive charge of the transition metal ion then increases and its bond with the olefin molecule becomes stronger.

Direct experimental proof of this suggestion has been obtained recently by Spiridonov and Kadushin in our laboratory. When oxygen-propylene mixtures are adsorbed on a reduced Mo-MgO catalyst, the intensity of the signal from O_2^- stabilized on Mg ions markedly increases, compared to that for individual adsorption of oxygen on this catalyst. Of the several forms of oxygen adsorption on Mg ions with $g_{||}$ values from 2.070 to 2.090, a great increase in intensity is observed only for the form with $g_{||} = 2.090$, the change in intensity of other forms being negligible. This might be evidence of a possible contribution from Mg atoms of hydrocarbon-oxygen mixtures having a certain positive charge. The charge value depends on the extent of screening of the Mg ions with OH groups.

When oxygen-hydrocarbon mixtures are adsorbed on CoO·MgO catalysts, the intensity of the O_2^- signal decreases.

According to the results obtained at the Institute of Chemical Physics, supported molybdenum oxide catalysts accelerate partial olefin oxidation at moderate Mg concentrations (1-5%), and the supported cobalt oxide catalysts accelerate complete oxidation of olefins.

The difference in the nature of electron transfer between adsorbed molecules may be evidence that complete and partial oxidation occur via different intermediate complexes. The mechanism of partial propylene oxidation[8] to acrolein with participation of the O_2^- ion radical proposed by us does not contradict the above results.

References

1) O. V. Krylov, Problemy Kinetika i Kataliza, 13, Kompleksoobrazovanie v Katalize, 141 (1968).

2) D. A. Dowden and D. Wells, in Actes du 2-me Congres International de Catalyse (1960), Paris, Technip, 2, 1489, 1961.

3) G. Asmolov and O. V. Krylov, Kinetika i Kataliz, 11, 1028 (1970).

4) G. N. Asmolov and O. V. Krylov, Izv. Akad. Nauk SSSR, Khim., No.10, 2414 (1970).

5) K. N. Spiridonov, G. B. Pariiskii and O. V. Krylov, Izv. Akad. Nauk SSSR, Khim., No.11, 2646 (1970).

6) G. N. Asmolov and O. V. Krylov, Kinetika i Kataliz, 12, 463, (1971).

7) J. H. Ashley and P. C. H. Mitchell, J. Chem. Soc., A,

1821 (1968).

8) O. V. Krylov and L. Ya. Margolis, Kinetika i Kataliz, 11, 432 (1970).

9) J. M. Lipsch and G. C. A. Schuit, J. Catalysis, 15, 163 (1969).

10) O. V. Krylov, G. B. Pariiskii and K. N. Spiridonov, J. Catalysis (1971), to be published.

11) P. H. Kasai, J. Chem. Phys., 44, 3325 (1965).

12) I. D. Mikheikin, A. I. Mashchenko and V. B. Kazanskii, Kinetika i Kataliz, 8, 1363 (1967).

13) J. H. Lunsford and J. P. Jayne, J. Chem. Phys., 44, 1387 (1966).

14) A. A. Gezalov, G. M. Zhabrova, V. V. Nikisha, G. B. Pariisii and K. N. Spiridonov, Kinetika i Kataliz, 9, 462 (1968).

15) K. I. Matveev, Kinetika i Kataliz, 10, 717 (1969).

16) L. H. Little, Infrared Spectra of Adsorbed Species, Academic Press, London - N. Y. - Paris, 1966.

17) J. M. Peacock, M. J. Sharp, A. L. Parker, P. G. Ashmore and J. M. Hockey, J. Catalysis, 15, 379 (1969).

18) A. L. Dent and R. J. Kokes, J. Am. Chem. Soc., 92, 6709 (1970).

Mechanism of Carbon Monoxide Oxidation on Manganese Dioxide Catalysts as Revealed by a Transient Response Method

H. Kobayashi and M. Kobayashi

Hokkaido University, Sapporo

INTRODUCTION

Steady state flow methods have frequently been used for kinetic studies of heterogeneous reactions. Because of the complexity of the reaction sequences on the catalyst, the information obtained merely from analysis of the reaction components under steady state conditions is insufficient to elucidate the mechanism of the reaction. To assess the validity of the results on the basis of purely mathematical criteria is an unsatisfactory procedure, although without further data it is the best that can be done. It is only recently that many instrumental techniques have been developed for direct analysis of the intermediates formed on the catalyst surface. These methods have provided much valuable information on heterogeneous catalysis and have proved to be indispensable. Unfortunately, however, there are some limitations in applying these methods to all types of reaction on various catalysts.

In order to get as much information as possible concerning the events on the catalyst surface, an attempt was made to introduce a transient response technique to the steady state flow method. It is based on the ordinary flow method, but the introduction of a minor purturbation can differentiate the rate events on the catalyst under any reaction conditions. Although further help would be needed to determine the rates of all the elementary steps, simulation with an electronic computor can be a useful tool for this purpose.

In the present study, the catalytic oxidation of carbon monoxide on manganese dioxide catalysts was studied by this method.

TRANSIENT RESPONSE METHOD

There are several different methods which can be applied
to follow the non-steady state of a reaction system. In the
present study, the transient response to a step change in the
concentration of reaction components was employed. A
schematic diagram of carbon monoxide oxidation on the
catalyst is shown in Fig. 1. A differential flow type reactor

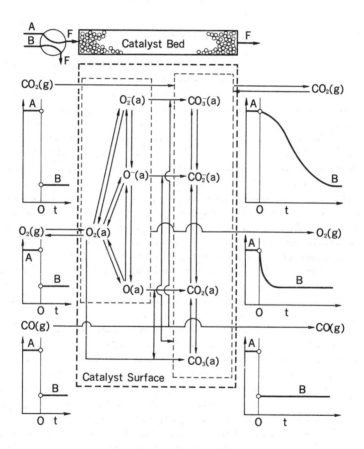

Fig. 1. Schematic diagram for the transient response
method.

with an appropriate amount of the catalyst was used, and the
system was kept under steady state conditions with a constant
total flow rate at given temperatures. By using the four way
valve attached immediately in front of the reactor, the reaction
gas mixture was switched over from one to a second in which
one component is present in a slightly different concentration.
Care was taken to keep the total flow rate as constant as
possible and, in addition, concentration jump was restricted
to a low value so that no appreciable temperature change was
caused. The response of the exit gas concentration and/or
electrical conductivity of the catalyst was followed by a gas
chromatograph and a conductivity measurement device.
The characteristic behavior of the response curve can
provide information on the amount and the state of adsorbed
species, the nature of the adsorption sites, and the rates of
several elementary steps including the charge transfer from
and to the catalyst.

EXPERIMENTAL

The catalyst employed was an electrolytic manganese
dioxide composed of 60 to 80 mesh granules. It was dried by
heating in a flowing gas mixture of 20 vol. % of oxygen and
80 vol. % of nitrogen at 110°C for 24 hrs. The BET surface
area (using nitrogen) was 30.8 m^2/g, and Hall effect measurement showed that this catalyst was an n-type semiconductor.
A schematic diagram of the system used is shown in
Fig. 2. The reactor consists of U tube of 0.55 cm I.D. and
83 cm length, packed with 39.8 g catalyst. It was immersed
in a methanol bath, the temperature of which was controlled
in a range from -5°C to -26°C with an accuracy of 0.05°C.
Total flow rate of the gas was kept constant at 180 + 1 mℓ/min.
Whenever the gas stream was switched, the change in the
total flow rate was within 1 %. After changing the inlet gas
composition, the outlet gas composition was analysed by gas
chromatography every 30 sec. for 5 min. At this flow rate,
the mean residence time of the gas in the reactor was 4 sec.
The intraparticle diffusion resistance of the catalyst was
found to be negligibly small by checking with rate data at 50°C.
The initial response of an He-N_2 mixture revealed that the
non-adsorptive response was completed within 10 sec. The
temperature of the catalyst bed was measured at both the inlet
and the outlet of the bed and ΔT was less than 0.1°C. The
total conversion was always less than 5%, which allows us to

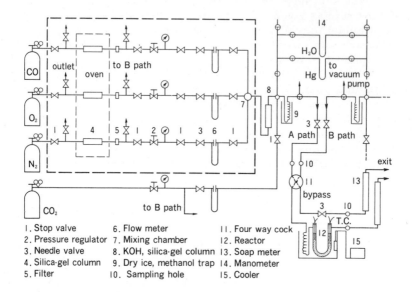

1. Stop valve
2. Pressure regulator
3. Needle valve
4. Silica-gel column
5. Filter
6. Flow meter
7. Mixing chamber
8. KOH, silica-gel column
9. Dry ice, methanol trap
10. Sampling hole
11. Four way cock
12. Reactor
13. Soap meter
14. Manometer
15. Cooler

Fig. 2. Schematic diagram of apparatus for transient response investigation.

consider this reactor to be effectively a differential reactor. Typical response curves are shown in Fig. 3, which demonstrates sufficient reproducibility of the data.

The conductivity of the catalyst was measured by using either D.C. or A.C. of various frequencies up to 30 kHz. Since the results obtained by both methods showed similar behavior, D.C. measurements were commonly used.

The surface oxygen species were analysed by the KI method, which was discussed in a previous paper[1]. In the present study, however, the method was improved so that the catalyst could be analysed without exposing it to air.

EXPERIMENTAL RESULTS AND DISCUSSIONS

The response of component B in the outlet gas mixture to a step change of the concentration of A in the inlet gas stream is designated as A-B response. When A is increased, this becomes A(inc.)-B, and when A is decreased, A(dec.)-B.

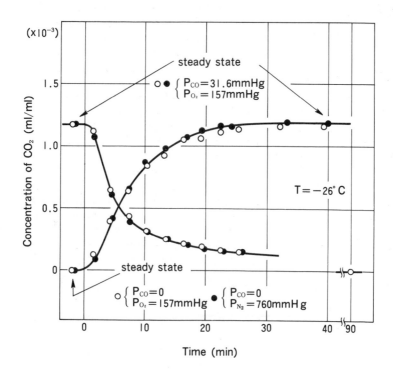

Fig. 3. Response of CO_2 concentration to stepwise change
in partial pressures of CO and O_2.

Behavior of Carbon Monoxide

It is essential in discussing the mechanism of this
reaction to determine whether the adsorption of CO on the
catalyst is significant.

The results for CO-CO response are shown in Fig. 4.
The response was instantaneous and this suggests that the
adsorption of CO, if it occurs at all, takes place at a very
slow rate or to a very minor extent. If CO is adsorbed on
the catalyst surface as an electron donor, the electrical
conductivity of the catalyst will increase with increasing
partial pressure of CO in the gas phase. The results of
conductivity measurements during steady state reaction under
various partial pressures of carbon monoxide indicated no

appreciable P_{CO}-dependency of the conductivity, as can be
seen from Fig. 5. The discontinuous change in conductivity
at $P_{CO} \cong 65$ mmHg can be attributed to a structural change in
the catalyst surface layer, as discussed in a previous paper[1].
The CO-CO response was also measured with a catalyst
which had been reduced for 48 hrs in a stream of carbon
monoxide-helium mixture, and the results again showed the
instantaneous response of CO with no appreciable change in
the conductivity, as seen in Fig. 6. The introduction of
oxygen to this catalyst produced no detectable amount of CO_2
and, moreover, no desorbed carbon monoxide was detected
even when this catalyst was heated up to 90°C in a helium
stream. No evidence was observed with catalysts which
were effective in the reaction gas mixture under steady state
conditions to show any adsorption of carbon monoxide on the
catalyst. Roginskii et al[2] and Brooks[3] among others, suggested a reaction between adsorbed carbon monoxide and
gaseous oxygen. The present results, however, indicate that

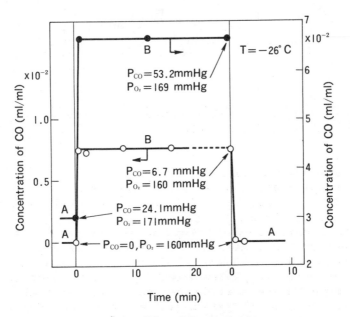

Fig. 4. CO - CO response.

no significant contribution from this reaction exists. As can
be seen from Fig. 3, the excellent agreement of the two
response curves of carbon dioxide to the switch over of inlet
gas mixtures, one containing oxygen and the other not, provides further evidence for the above interpretation. It is
reasonable to conclude from these results that carbon
monoxide will react from the gas phase with the oxygen
species on the surface.

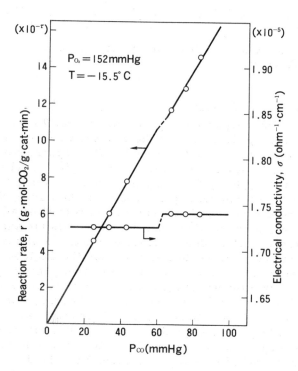

Fig. 5. Electrical conductivity and reaction rate as a
function of P_{CO}.

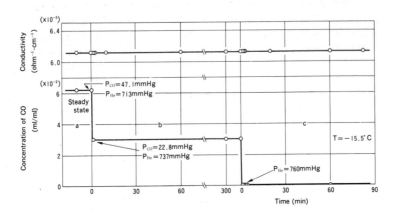

Fig. 6. CO - CO and CO - conductivity response on a
reduced catalyst.

Behavior of Carbon Dioxide

The CO_2-CO_2 response curves shown in Fig. 7 clearly
indicate the adsorption of CO_2 on the catalyst surface. The
graphical integration of the CO_2 (inc.) -CO_2 response curve
gives the amounts of CO_2 adsorbed under the given conditions
and that of the CO_2(dec.)-CO_2 response gives the amounts of
desorbed CO_2. As can readily be seen from the figure, these
amounts were not equal and this fact suggests that the adsorption of CO_2 consists of two parts, one reversible and
the other irreversible. The reversible amounts of adsorbed
CO_2 at -26 and -21°C were determined from the CO_2(dec.)-CO_2
response curves measured in various gas atmospheres by
shifting P_{CO_2} over different ranges, and the results are shown
in Fig. 8. The amounts of reversibly adsorbed CO_2 were the
same regardless of the different gas mixtures and varied only
with P_{CO_2} according to a Langmuir isotherm. The saturated
amount was found to be $6.43 \times 10^{13} CO_2/cm^2$. The amount of
irreversibly adsorbed CO_2 was calculated from the difference
between the amounts obtained from both curves and were
found to be in a range $2 \sim 6 \times 10^{12} CO_2/cm^2$ depending upon
P_{CO_2}. Elovitch[4] has observed that a fractional amount of CO_2
adsorbed on manganese dioxide was very strongly adsorbed,
and Klier and Kuchynka[5] have also found the amount of irreversibly adsorbed CO_2 at 0° and 20°C to be about 6×10^{12}
CO_2/cm^2. The present results are in good agreement.

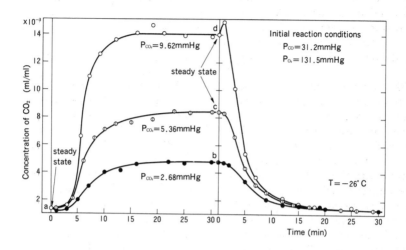

Fig. 7. CO_2 - CO_2 response during the reaction.

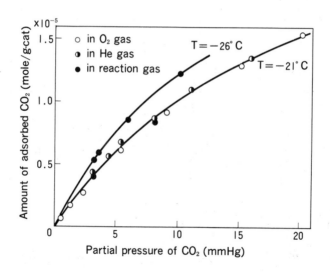

Fig. 8. Adsorption isotherm for CO_2.

When a jump in the partial pressure of carbon dioxide
was imposed on the inlet gas stream during the reaction, the
amount of adsorbed CO_2 varied depending upon its partial
pressure in the gas phase, as stated above, but the electrical
conductivity of the catalyst and the rate of carbon monoxide
oxidation on the catalyst were not affected by this. Roginskii
et al.[2] have also observed that the existence of CO_2 had no
effect on the rate of this reaction. These results suggest
that there is no electron transfer accompanying the adsorption and desorption of CO_2 on the catalyst, and also that the
sites on which this reversible CO_2 is adsorbed are different
from the sites on which the active species of surface oxygen
exist. As will be discussed later, the catalytically active
oxygen on the surface, O_s^h , is analysed by the KI method, and
the concentration of O_s^h on the surface was observed not to be
affected by the existence of CO_2. This fact also provides
further support for the above interpretation.

The CO_2-CO_2 response on a catalyst which had been
reduced by a carbon monoxide-helium mixture for 48 hr gave
the curves shown in Fig. 9, which are quite similar to that in

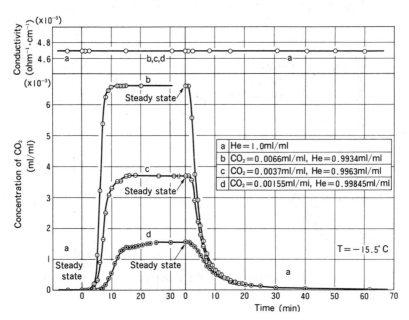

Fig. 9. CO_2 - CO_2 and CO_2 - conductivity responses on a
reduced catalyst.

Fig. 7. The electrical conductivity of the catalyst did not
exhibit even a slight change in this case and the amount of
reversibly adsorbed CO_2 was well correlated by a Langmuir
isotherm. This fact, together with the observations shown
in Fig. 8, strongly suggest that the adsorption sites for the
reversible CO_2 are Mn ions on the surface. The heat of
adsorption of CO_2 was found to be 7.6 kcal/mole in this experiment, which is slightly higher than the value of 5.5
kcal/mole obtained by Roginskii and Zel'dovitch.[6]

As noted in the previous section, there are no appreciable
amounts of adsorbed CO on the surface. Hence, it does not
seem implausible to conclude that CO_2 is initially formed on
the active surface oxygen O_s^h by reaction with gaseous CO
followed by transfer to different sites, Mn ions, before
desorption from the surface.

Behavior of Oxygen
The O_2-O_2 response during the steady state reaction
clearly showed a delay which indicates the adsorption of
oxygen on the catalyst surface, as shown in Fig. 10. In
order to examine further the nature and amount of oxygen on
the surface, the reactions between gaseous CO and oxygen on
the surface were followed under various conditions. Results
obtained are presented in Fig. 11.

Catalyst used for the steady state reaction was subjected
to treatment with oxygen-helium mixtures of various concentrations (tabulated in the figure). After replacement with
a pure helium stream for 30 min, this stream was switched
over to helium streams containing carbon monoxide at
various different partial pressures as indicated along the
curves. This procedure was repeated as indicated by the
numbers along the lines. The first response is shown by
curve (1). The second response (2) traced almost the same
curve as (1) because the oxygen consumed by the previous
reaction was completely replaced by subsequent oxidation
treatment with a sufficiently high concentration of oxygen for
48 hr. The third response was found to be much lower than
the preceeding ones and this may be attributed to the shorter
period of preoxidation. After this, the response was followed
until ultimately the change in CO_2 concentration was too small
to be detected, then the catalyst was reoxidized and the response was again followed. Despite the difference in the
partial pressure of oxygen in the oxidizing gas mixture, the
response data traced the curve designated by (4), although

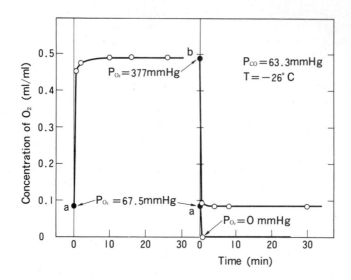

Fig. 10. $O_2 - O_2$ response during the reaction.

the previous response curve was followed until it ultimately reached a quasi-steady state. This response was found even after a much shorter period of oxidation, say for 50 min.

These results suggest that the active surface oxygen is regenerated in two different ways, one being fairly rapid and the other slow. Since there was no oxygen in the gas phase, the finite and constant rates of CO_2 formation in the last parts of responses (3) and (4) further suggests that there exists some amount of less active oxygen which slowly transforms into active oxygen. The amount of active oxygen on the surface calculated from curve (4) was about $2.5 \times 10^{13} O_2/cm^2$ and the amount of less active oxygen was found from curve (1) to be $9.5 \times 10^{13} O_2/cm^2$.

These conclusions are further supported by the following results, obtained by the analysis of surface oxygen by means of the KI method. As can be seen from Fig. 12, a fresh catalyst has its surface oxygen distributed into two parts with respect to relative oxidation power; oxygen in the region of lower oxidation power, O_s^l, and oxygen in the region of higher oxidation power, O_s^h. After reduction with CO for 48

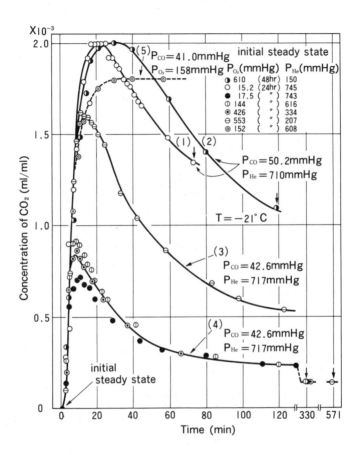

Fig. 11. Transient response of CO_2 concentration to stepwise change in partial pressures of CO and O_2.

hr O_s^h vanished but O_s^ℓ still remained. Reoxidation with oxygen for 30 min. fully regenerated the O_s^h. The amounts of O_s^h and O_s^ℓ analysed by the KI method were in a range $0 \sim 9.4 \times 10^{13} O_2/cm^2$ (depending upon catalytic activity, which varied with the introduction of moisture) and $1.0 \sim 1.7 \times 10^{14} O_2/cm^2$ (depending upon P_{CO}), respectively. These value are very

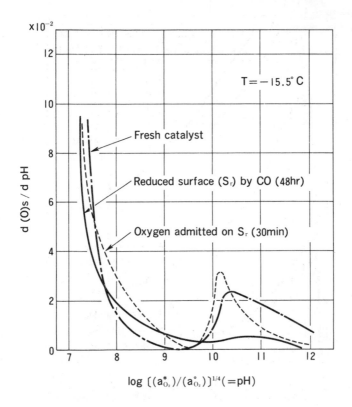

Fig. 12. Distribution curves of oxidation power for the surface oxygen of MnO_2.

similar to those obtained from the response studies. The events shown by (1) ∿ (4) in Fig. 11, therefore, may reasonably be considered to represent reaction between gaseous carbon monoxide and O_s^h on the surface.

The rate of carbon monoxide oxidation on MnO_2 catalyst is first order with respect to P_{CO}. The rate constants were found to be proportional to the amounts of O_s^h as shown in Fig. 13. This fact further supports the above conclusions.

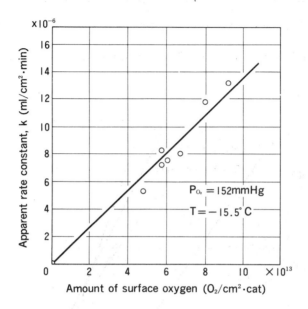

Fig. 13. The apparent rate constant vs. amount of O_s^h.

The curve (5) in Fig. 11 represents the CO_2 response to the introduction of a CO and O_2 mixture. The initial rise in the response curve is exactly the same as that of other response curves which were obtained when only CO without O_2 was introduced. This suggests that only O_s^h and not gaseous oxygen, plays the major role in this reaction.

It is of interest to know whether O_s^h is neutral or ionized. In order to investigate this problem, it will be worthwhile to examine the KI method in detail. In the KI method, the reduction of surface oxygen was considered to take place by the coupling of the following two simultaneous reactions,

$$\tfrac{1}{2} O_2 + 2H^+ + 2e = H_2O \qquad\qquad (1)$$

$$I_3^- + 2e = 3I^- \qquad\qquad (2)$$

According to our recent measurements on the equilibrium potential of reaction (2) on an MnO_2 electrode, it was found that the equilibrium I_3^-/I^- system passed into an IO_3^-/I^- system at pH values higher than 11.25, as shown in Fig. 14.

$$IO_3^- + 6H^+ + 6e = I^- + 3H_2O \tag{3}$$

The equilibrium potentials of these reactions are given as follows.

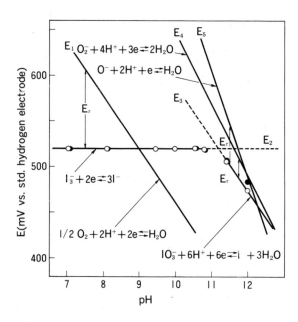

Fig. 14. Diagram of oxidation-reduction potential as a function of pH:O, experimental data:$I_3^-/I^- = 0.0126$.

$$E_1 = E_1^0 + (RT/2F) \ln (a_{O_2})(a_{H^+})^2/(a_{H_2O})$$

$$E_2 = E_2^0 + (RT/2F) \ln (a_{I_3^-})/(a_{I^-})^3$$

$$E_3 = E_3^0 + (RT/6F) \ln (a_{IO_3^-})(a_{H^+})^6/(a_{I^-})(a_{H_2O})^3$$

$$E_4 = E_4^0 + (RT/3F) \ln (a_{O_2^-})(a_{H^+})^4/(a_{H_2O})^2$$

$$E_5 = E_5^0 + (RT/F) \ln (a_{O^-})^2(a_{H^+})^2/(a_{H_2O})$$

$$E_1 = E_1^0 + \frac{RT}{2F} \ln \frac{(a_{O_2})^{\frac{1}{2}}(a_{H^+})^2}{(a_{H_2O})} \qquad (1')$$

$$E_2 = E_2^0 + \frac{RT}{2F} \ln \frac{(a_{I_3^-})}{(a_{I^-})^3} \qquad (2')$$

$$E_3 = E_3^0 + \frac{RT}{6F} \ln \frac{(a_{IO_3^-})(a_{H^+})^6}{(a_{I^-})(a_{H_2O})^3} \qquad (3')$$

The pH dependencies of E_1, E_2 and E_3 are 59.1, 0 and 59.1 mV/pH, respectively. At pH values lower than 11.25, reaction (1) will be coupled with reaction (2) and the electromotive force E_r for the reduction of surface oxygen by I^- ion will therefore be given by $E_r = E_1 - E_2$. Since this E_r is dependent on pH, the distribution of oxidation power of oxygen species can be obtained by measurement at different pH values. At pH values higher than 11.25, if reaction (1) is coupled with reaction (3), then E_r, which is given by $E_1 - E_3$, is a constant independent of pH, since both E_1 and E_3 have the same pH dependencies, and therefore the amount of surface oxygen measured in the higher pH region should be a constant independent of pH. As can be seen in Fig. 12, this was not the case, and the distribution of surface oxygen was dependent upon pH even in this higher pH region. This suggests that the reaction coupled with reaction (3) is not reaction (1), but one or both of the following reactions.

$$O_2^- + 4H^+ + 3e = 2H_2O \qquad (4)$$

$$O^- + 2H^+ + e = H_2O \qquad (5)$$

The equilibrium potentials of these reactions are given as follows

$$E_4 = E_4^0 + \frac{RT}{3F} \ln \frac{(a_{O_2})(a_{H^+})^4}{(a_{H_2O})^2} \qquad (4')$$

$$E_5 = E_5^0 + \frac{RT}{F} \ln \frac{(a_{O^-})(a_{H^+})^2}{(a_{H_2O})} \qquad (5')$$

Since E_4 and E_5 have pH dependencies of 78.8 and 118.2 mV/pH, respectively, the E_r of a coupled system of reaction (3) with either reaction (4) or reaction (5) will also be

dependent on pH, and the distribution of oxidation power of
the surface oxygen can be measured as shown in Fig. 12.
Based on these discussions, threfore, we may reasonably
assume that the catalytically active oxygen O_s^h measured
in the higher pH region is either or both of the two types
of oxygen ions, i. e. , O_s^- and O^-.

In this regard, the following response will provide
valuable information. When the carbon monoxide oxidation
reaction was taking place under steady state conditions, the
stream of reaction gas mixture was switched over to a
stream of pure helium and the change in both the electrical
conducitivity and the concentration of CO_2 formed were
followed simultaneously. The results are shown in Fig. 15.
The electrical conductivity decreased gradually until it
reached a new steady state after 20 min. The concentration
of released CO_2 also decreased gradually until ultimately no
appreciable amount was detected after 25 min.

Bearing in mind that no conductivity change was observed
when CO_2 was desorbed from the surface, we may reasonably
conclude that the decrease in conductivity can be attributed to
the pick up of an electron from the catalyst by electrically
neutral oxygen, presumably an oxygen molecule, which had
been adsorbed on the surface, since there was no oxygen in
the gas phase.

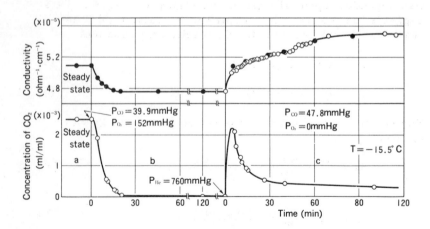

Fig. 15. Response of conductivity and CO_2 concentration to
step change of gas composition.

After keeping this catalyst in the helium stream for
several hours, the gas inflow was again switched over to a
mixture of helium and carbon monoxide. The electrical
conductivity again increased due to the release of electrons
according to the following successive reactions,

$$O_2^-(s_1) + CO\ (g) \longrightarrow CO_3^-(s_1) \tag{6}$$

$$CO_3^-(s_1) + s_2 \longrightarrow O^-(s_1) + CO_2(s_2) \tag{7}$$

$$O^-(s_1) + CO\ (g) \longrightarrow CO_2^-(s_1) \tag{8}$$

$$CO_2^-(s_1) + s_2 \longrightarrow e + s_1 + CO_2(s_2) \tag{9}$$

where s_1 and s_2 denote the different adsorption sites.
If the catalytically active oxygen species consists of O^- ions
alone, the over-all reaction will proceed only through
reactions (8) and (9), but if O_2^- ions exist accompanied by O^-
ions, the over-all reaction must proceed through all four of
these reactions.

The variation in the conductivity of the catalyst during
these responses again revealed that the catalytically active
oxygen species on the surface are present as ions and also
suggested the important role of electrons in this reaction.
There is no evidence leading us to prefer either O_2^- or O^-
as the catalytically active oxygen species on the surface.
At low temperatures such as were used in this study, however,
it would be reasonable to assume that O_2^- ions are initially
formed on the surface and O^- ions are then formed by reactions
(6) and (7). Winter[7] has suggested that oxygen adsorbed on
nickel oxides at temperatures higher than room temperature
up to 250°C is in the form of O_2^-, but he favored electrically
neutral oxygen molecules at temperatures lower than room
temperature. In the case of manganese dioxide, as we found,
both ionized and electrically neutral oxygen exist on the
surface but the former is active and plays the major role in
the reaction.

MECHANISM OF THE REACTION

According to the results and discussions presented so far,
the oxidation of carbon monoxide on manganese dioxide
catalysts appears to take place as a sequence of the following
steps.

$$O_2(g) + e + s_1 \longrightarrow O_2^-(s_1) \tag{10}$$

$$O_2^-(s_1) + CO(g) \longrightarrow CO_3^-(s_1) \tag{6}$$

$$CO_3^-(s_1) + s_2 \longrightarrow O^-(s_1) + CO_2(s_2) \tag{7}$$

$$O^-(s_1) + CO(g) \longrightarrow CO_2^-(s_1) \tag{8}$$

$$CO_2(s_1) + s_2 \longrightarrow s_1 + e + CO_2(s_2) \tag{9}$$

$$CO_2(s_2) \overset{\longrightarrow}{\longleftarrow} CO_2(g) + s_2 \tag{11}$$

Besides these main reactions, oxygen molecules which are adsorbed on the sites s_2 will pick up electrons very slowly and thus take part in the reaction. However, the contribution to the reaction through this path is small compared with that of the main path presented above.

The irreversibility of reaction (10) is evident from the fact that a decrease in the amount of O_s^h could not be detected when the catalyst was kept in a stream of helium for various periods of time. Reactions (6) and (8) can also be regarded as taking place irreversibly, because carbon monoxide could not be detected when catalyst which had been reacting under steady state conditions was heated in a helium stream. Reactions (7) and (9) can also be considered to be irreversible because there were no detectable effects of carbon dioxide on the over-all reaction rate. Only the desorption of carbon dioxide (11) is reversible.

Since the overall reaction rate is first-order with respect to the partial pressure of carbon monoxide, the rate of either reaction (6) or (8) will be the slowest. Since, also, the electrical conductivity of the catalyst remains unchanged regardless of the over-all reaction rate, as demonstrated in Fig. 5, reaction (10) can proceed rapidly only when electrons are released by reaction (9) and this may be the reason why the amount of O_s^h is constant regardless of the over-all reaction rate, as demonstrated in a previous paper[1] That the reaction rate is independent of the partial pressure of oxygen may also be ascribed to the same reason.

ANALYSIS OF THE REACTION RATE

Because of the facts that the over-all reaction rate is

first-order with respect to the partial pressure of carbon monoxide and that the amount of O_s^h is constant regardless of the over-all reaction rate, it is reasonable to assume that the rates of reactions (7) and (9) are rapid compared with those of reactions (6) and (8), respectively. In order to simplify kinetic analysis, therefore, the over-all reaction can be written as follows.

$$O_2(g) + e + s_1 \xrightarrow{k_1} O_2^-(s_1) \tag{10}$$

$$O_2^-(s_1) + CO(g) + s_2 \xrightarrow{k_2} O^-(s_1) + CO_2(s_2) \tag{12}$$

$$O^-(s_1) + CO(g) + s_2 \xrightarrow{k_3} s_1 + e + CO_2(s_2) \tag{13}$$

$$CO_2(s_2) \underset{k_5}{\overset{k_4}{\rightleftharpoons}} CO_2(g) + s_2 \tag{11}$$

From experimental data obtained for the adsorption of carbon dioxide on the catalyst during reaction at -26 °C, it was found that the adsorption equilibrium constant of carbon dioxide $K_{CO_2} = 46$, i.e., $k_5 = 46k_4$.

According to the reaction scheme presented in Eq. (10), (12), (13) and (11), the unknown parameters k_1, k_2, k_3 and k_4 were determined by simulation on a digital computer using Marquardt's method.[8-10] The experimental data for CO-CO$_2$ response were used for this purpose. The values of rate constants for each step determined by this method are;

$$k_1 = 5.46 \times 10^{-3} \, \text{mole}/\text{g} \cdot \text{min} \cdot \text{atm}$$

$$k_2 = 4 \times 10^{-6} \, \text{mole}/\text{g} \cdot \text{min} \cdot \text{atm}$$

$$k_3 = 2 \times 10^{-4} \, \text{mole}/\text{g} \cdot \text{min} \cdot \text{atm}$$

$$k_4 = 9.8 \times 10^{-5} \, \text{mole}/\text{g} \cdot \text{min}$$

$$k_5 = 4.5 \times 10^{-3} \, \text{mole}/\text{g} \cdot \text{min} \cdot \text{atm}$$

The fraction of s_1 occupied by O_2^-, θ_1, was found to be 0.98 and that by O^-, θ_1', was 0.02. The calculated curve for CO-CO response using these values was in satisfactory agreement with the curve obtained by experiment. These results are presented in Fig. 16.

By inspecting the values of the rate constants, we can see that the reaction between gaseous carbon monoxide and

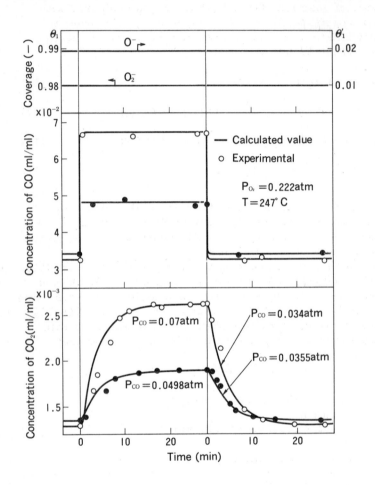

Fig. 16. CO - (θ_1, θ_1'), CO -CO and CO - CO_2 response.

surface oxygen O_2^- is the slowest, and the desorption of carbon dioxide from the surface is the second slowest step. The sum of θ_1 and θ_1' is unity and this means that the surface sites s_1 for active oxygen species are fully occupied under the reaction conditions. These results, obtained by kinetic analysis of transient response data, are in good agreement with experimental findings described so far.
Conclusion

CONCLUSION

A transient response technique was successfully applied to the catalytic oxidation of carbon monoxide on manganese dioxide catalysts. Valuable information on the behavior of the reaction components was obtained by analysis of the response curves. Based on this information, a detailed reaction mechanism was proposed and the kinetics of the overall reaction were analyzed and discussed. The transient response technique was shown to be a useful method for the study of heterogeneous catalytic reactions.

References

1) M. Kobayashi, H. Matsumoto, and H. Kobayashi, J. Catalysis, 21, 48 (1971).
2) M. Katz, Advan. Catalysis, 5, 179 (1953).
3) C. S. Brooks, J. Catalysis, 8, 272 (1967).
4) S. J. Elovitch, in Sec. Int. Cong. Surface Activity II, p. 252, London (1957).
5) K. Klier and K. Kuchynka, J. Catalysis, 6, 62 (1966).
6) S. Roginskii and Ya. Zel'dovich, Zhur. Fiz. Khim., 1, 554 (1934).
7) E. R. S. Winter, J. Catalysis, 6, 35 (1966).
8) D. W. Marquardt, J. Soc. Ind. Appl. Math., 11, 431 (1963).
9) W. E. Ball and L. C. D. Groenweghe, I & EC Fundamentals, 5, No. 2, 181 (1966).
10) D. M. Himmelblau, Process Analysis By Statiatical Methods, John Wiley & Sons, New York, 1970.

Discussion

V.F. Anufrienko

I would like to note one peculiarity of Professor Kwan's report; namely, that in this work the electron state of the Fe^{3+}ion in a complexes changed by axial coordination with nitrogen containing bases capable of forming a coordination bond with Fe^{3+} ion. The relationship between the changes of the Fe^{3+} ion state at coordination and the properties of bases is not discussed in the report. The clarification of such properties seems to throw light on the subsequent capacity of a complex to interact with oxygen. At the Institute of Catalysis the method of studying the peculiarities of coordinated and noncoordinated states (and transitions between them) of a complex was developed for the case of cuprous complexes (Cu^{2+}), (Zsurnal Strukturnoi Khimii, 12 No. 4, 1971). It proved to be possible to relate the energy of transition states into each other with the properties of bases and complexes. It seems possible to use this method in studying the coordination interaction of Fe^{3+}complexes being discussed in Professor Kwan's work.

V.B. Kazanski

Professor Kwan did not observe the formation of oxygen radicals in the case of iron-containing phthalocyanins although they were observed by the EPR method for phthalocyanins containing Co (II). I would like to offer the explanation for this fact. The case is that the spectrum of EPR radicals of O_2^- can be observed only if they are situated in a coordinate sphere of a diamagnetic ion. If the central ion of a complex is paramagnetic, the signal can be greatly widened and is not observed. This condition corresponds to the case of low spin bivalent cobalt, as before the interaction with oxygen its spin is equal to $1/2$ and after the interaction with O_2 it is equal to zero. In the case of iron, after the transfer of an electron Fe^{3+}, paramagnetic states are formed in the case of a low spin (s =½) as well as in the case of a high spin (s = ⅝) complexes. Therefore, it is possible that the spectra of EPR radicals of O_2^- in this case are not observed.

In connection with Professor O.V. Krylov's report I

would like to note the following: In the catalysts studied by him which were supported on MgO and Al_2O_3, the formation of solid solutions can take place. Therefore, it would be desirable to have data concerning the portion of Co and Mo ions present on the surface as well as in the volume of a catalyst. Such types of information can be obtained while studying the influence of the adsorption of polar molecules on optical and EPR spectra. In this case, for the surface ions the measurement of spectrum parameters should take place while the bulk ions inside of a catalyst are evidently insensitive to adsorption. As long as this is not done it is impossible to speak of the study of a coordination of surface ions, and it is possible that the experimental results refer to the formation of volume phases.

A. Matsuda

In the system of platinum and nickel hydrogen electrodes in alkali solutions the electrode potential can be divided into two independent parts which are described by the following equations:

$$Y_{10} = \frac{RT}{F} \ln A_{M^+} + \text{const.}$$

$$Y_{20} = -\frac{RT}{F} \ln A_M + \text{const.}$$

where Y_{10} is the electrode potential caused by free charges on the electrode surface and Y_{20} the potential caused by the accumulation of intermediate alkali-metal atoms on the electrode surface or in the bulk of the electrode metal near the surface. In the case when the alkali metal atom is adsorbed only on the electrode surface the latter equation shows the Frumkin-Temkin isotherm

$$\theta = \frac{1}{f} \ln a_M$$

where f has a clear physical sense and is expressed as

$$f = 4\Pi A \mu F / RT.$$

It will be very interesting to compare the constant F obtained by us and theoretically calculated by Professor Temkin on the basis of a two-dimensional free electron gas

model, although the latter model is too simple to explain
various types of adsorption and bond energies.

The model given by Professor Temkin is also interesting from the kinetic point of view in electrochemcial reactions,
as it permits us to explain the Tafel constant by the change
of the electron effective mass in a metal.

In the case when the alkali metal atom penetrates into
the bulk of the electrode metal the equation for Y_{20} shows the
contact potential difference between the surface compound
layers in the bulk of the electrode metal.

I would like to emphasize that this equation permits us
to combine the results obtained in a vacuum system with the
setting up of the electrode potential in the metal-solution
system.

The electron work function, however, was usually
measured in a vacuum system as a function of surface
coverage with adsorbed atoms. It is desirable to measure
the work function as a function of the activity of adsorbed
atoms for the explanation of the electrode potential on the
basis of the electron work function.

I would like to point out that the electrode potential
caused by the change of the electrode metal work function
due to the formation of an intermetallic compound controls
the activity of an adsorbed hydrogen atom in the elementary
step of water molecule decomposition.

A.E. Shylov

On the basis of very low values of the activation energy
of D_2 - H_2 exchange Professor M.I. Temkin has drawn a
conclusion about the ionized state of H on the surface, as
even a free atom of H reacts with H_2 with relatively high
activation energy.

However, it should be taken into account that in the
coordination sphere of metals, reactions of ligands often
proceed much more easily than the corresponding gas phase
reactions of molecules with free radicals or atoms.

Low activation energy values may denote that surface
coordination compounds take part in the reaction, ionic state
of ligands (e.g., of H atoms) being unnecessary.

Professor Tamaru in his speech gave very interesting
and important data on new heterogeneous catalysts for ammonia synthesis. It is of particular interest that carbon
monoxide is not a reaction inhibitor. We are looking
forward to the further development of this work.

K. Tamaru
    In connection to Professor Kobayashi's report on the
transient response method and the reports by Professors
Tretyakov and Savchenko I would like to present briefly my
approach to the elucidation of the reaction mechanism by
way of the more direct route - a spectroscopic method.
It permits us to study the dynamic behaviour of chemisorbed
species under operating conditions and to determine real
reaction intermediates.
    Although there are many works on the spectroscopic
study of adsorbed species on a catalyst surface, most of
them describe only the nature of surface compounds. Even
if the adsorbed species are present on the catalyst surface
during the reaction course, from this it does not follow that
they are real reaction intermediates.
    One of the approaches to elucidation of the true mechanism
is a dynamic method of studying the behaviour of chemisorbed
species under their operating conditions when the reaction
conditions are rapidly changed. It can be achieved, for
instance, by the replacement of one reactant or product with
its isotopic species or by the change of the reaction temperatures or pressures of each type of species.
    In the course of the reaction, chemisorbed species are
constantly studied on the catalyst surface with the help of
spectroscopic methods to determine the kinds of chemisorbed species on the catalyst surface directly during the
reaction, their structures, their surface concentrations, and
simultaneously, the rate of the overall reaction.
    These observations were carried out under various
reaction conditions to examine the dependence of the reaction
rate on the amounts of each kind of chemisorbed species on
the catalyst surface as well as on the amounts of reactants
and ambient gas. Then this dependence is associated with
the chemical species which participate in the rate determining
step of the reaction.
    By such methods we studied the decomposition of formic
acid on alumina and ZnO, the reaction of water-gas shift on
ZnO, H_2 - D_2 exchange, and the alcohol decomposition.
As I am short of time I will show you only one example:
the alcohol decomposition on ZnO.
    At the contact of methyl alcohol vapors with ZnO, ions
of methoxide and formate were observed and molecules of
D_2, CO_2 and CO were evolved into a gas phase. At the re-

moval of CD_3OD from the ambient gas by a dry ice-methyl
alcohol trap during the reaction course the evolution of D_2
and CO_2 stopped while the evolution of CO continued unchanged. At 240°C the rate of surface formate decomposition was
in reasonably close agreement with the rate of CO formation.
During the release of the trapped methanol the rates of CO_2
and D_2 formation increased again, surface methoxide reap-
peared, and the concentration of the surface formate ion
decreased respectively. These results lead to the conclusion
that CO is formed mainly through the formate ion decomposition and D_2 and CO_2 came from the reaction between CD_3OD
(g) and DCOO (ads.).

L. Ya. Margolis
      In Professor Kobayashi's report it was shown that at
temperatures from -26°C to -5°C on MnO_2 the adsorption of
CO did not take place and the reaction followd the collision
mechanism with chemisorbed oxygen. Elovitch and I investigated CO oxidation on electrolytic MnO_2 and found that at
20°C carbon oxide was chemisorbed and interacted with
chemisorbed oxygen. At higher temperature, according
to the data of Bruns, the stage reaction of the conversion
of CO to CO_2 proceeds. Thus, depending on the
temperature range within which the reaction proceeds its
mechanism changes. The measurement of MnO_2 electnon
work function at CO adsorption showed that the adsorbed
molecule of CO was not charged.

G.K. Boreskov
      I would like to direct a question to Professor Kobayashi:
Do you insist upon the scheme of the reaction proceeding
through molecular oxygen at higher temperatures?

H. Kobayashi
      The experiments were made at low temperatures. At
high temperatures we expect the same mechanism.

G.K. Boreskov
      In the works of Bruns the stationary rate of CO oxidation
on MnO_2 was studied. It was shown that the rate of the stages
was coincidental with the rate of a catalytic reaction. It is
unlikely that this result can be correlated with the assumption
on the molecular oxygen participation unless it is thought
that molecular oxygen covers the greater part of the surface.

H. Kobayashi

It was found that the greater part of the surface was covered with oxygen, but only a small part of oxygen was active (10+25%) depending on the reaction conditions.

G.K. Boreskov

I highly appreciate the method itself that was given by Professor Kobayashi. I believe that this is the only method of finding intermediate products and of measuring nonstationary rates. But I am not sure that the reaction mechanism given unambiguously results from the experimental data obtained.

S.L. Kiperman

I would like to say a few words on the extremely interesting investigations with the use of ultravacuum reported here and published earlier. They do not deal with one significant question - whether the use of ultravacuum itself significantly changes the laws of catalysis on metals. The case is that in the above-mentioned works the precise correlation between the kinetic rules of reactions and their rates under indentical conditions was not performed. It is particularly important as ultravacuum treatment is always followed by other operations - thermic treatment of the catalyst and overall system. At our laboratory Dr. M.S. Kharson performed a special investigation in which he thoroughly studied the kinetics and rate of reactions of every step (other conditions being the same) - before treatment, after thermic treatment of the catalyst, after thermic treatment of the overall system, and after the action of ultravacuum of the order of 10^{-10} torr. Three reactions were studied: formic acid decomposition, carbon dioxide hydratation, and isotopic exchange of hydrogen with deuterium on nickel. At all stages the same kinetic equation and activation energy values were obtained for each reaction.

The reaction rates significally increased after the thermic treatment, but then they changed slightly. Thus, the rate of isotopic exchange of hydrogen after the thermic treatment of the catalyst at 860°C and of the system at 450°C increased by a factor of 4.8, and after treatment in ultravacuum, by a factor of 5.6. Therefore, in this case the most significant influence produces not the ultravacuum action itself but the thermic treatment eliminating impurities.

In connection with this one can note that, at least for the
systems studied by us, the threat of the alteration of true
catalysis laws on metals due to the "adsorption cover"
assumed by many authors and, in particular, by Professor
Roginski proves to be not so dangerous.

O.V. Krylov

Concerning the communication by Tretyakov and relating
to the speech by Kiperman, I should like to say the following.
The data of I.I. Tretyakov, contrary to the agreements of
Kiperman, show a principal difference between the data on
metals obtained in ultravacuum at a high degree of cleaning
and the data on metals purified by a usual technique. For
such metals as Pt or Pd and for oxidation of H_2 or CO the
reaction proceeds almost at every collision (e.g., of
molecules of H_2 or CO with an adsorbed oxygen layer)
practically without any activation energy. The reaction
proceeds even at -195°C. The calculated elementary rate
constant of interaction of CO with an adsorbed layer corresponds to the elementary rate constant of a reaction in a
gas phase of a molecule with very active free radicals.
The investigation aim is not to deduce a kinetic equation as
Kiperman believes but to obtain maximum possible absolute
rates. If the equation is the same for clean and impure
metals it is very well, as it will permit the rate constants
to be directly compared.

Concerning Professor Kobayashi's report it seems to me
that the response technique is much more convenient for the
investigation of nonstationary processes than, for instance,
a pulse method. Having used the response technique (our
attention was drawn to it by Professor Kobayashi) we examined
the reaction of ethanol-acrolein interaction with the production
of allyl alcohol on an oxide-magnesium catalyst. By such a
method we could, for instance, show that in a steady state
the number of active centers was two orders lower than in
the initial catalyst state.

In relation to V.B. Kazanski's speech I would like to
say that a part of ions of transition metals (Mo, V, Co, etc.)
deposited on a matrix of MgO or Al_2O_3 enters into a solid
solution, the other part remains on the surface. These ions
differ by their valence state and coordination as shown with
the help of EPR methods and diffuse reflection spectra.
This difference was proved by us experimentally. The
changes during adsorption were also taken into account. A

part of the signals at adsorption changed and the other part remained unchanged.

M.I. Temkin

In connection with the discussion which took place here concerning the importance of ultrahigh vacuum treatment for the investigation of the mechanism and kinetics of heterocatalytic reactions I shall mention the work of Langmuir published 50 years ago (I. Langmuir, Trans. Faraday Soc., 1921-1922). Having investigated the kinetics of the reaction of Co with O_2 on platinum, Langmuir came to the conclusion on the reaction mechanism reasonably close to those made by I.I. Tretyakov. Both investigators found that the reaction proceeded at the collision of a CO molecule out of a gas phase with an adsorbed atom of O, and that the adsorbed molecules of CO could not react with oxygen.

T. Kwan

One of the disadvantages of the EPR method is that the absence of an EPR spectrum does not necessarily mean the absence of paramagnetic species. We could not observe O_2^- in the case of diimidazole Fe porphyrin at 77°K while the paramagnetic anion was present in the case of low spin Co porphyrin. The reason for this discrepancy may be that as proposed by Professor Kazanski, $s = \frac{5}{2}$ or $\frac{1}{2}$ for Fe^{3+} and $s = 0$ for Co^{3+}. Unfortunately, the position of oxygen species relative to Fe^{3+} is unknown. The information obtained by different physical methods was taken into account in the present paper although a part of it is of a qualitative nature. I hope that the electron transfer mechanism proposed here will be critically investigated on the basis of quantitative measurements of various kinds.

H. Kobayashi

The response technique can provide extensive information which would be a sound basis for elucidating a reaction mechanism as well as for the kinetic description of heterogeneous catalysis. For the elucidation of the mechanism of a catalytic reaction it is very important to know about the intermediates on the surface. For this purpose many instrumental methods were used. But not all species on the surface take part in the main reaction, and it can be important to identify real intermediates among the species on the surface. In this case, as was shown in the present work

and in Professor Tamaru's speech, the response technique
or any other dynamic method can be quite useful. The results
of the simulation by a digital computer were quite satisfactory,
but the use of hybrid computers for such a purpose would be
more suitable.

I.I. Tretyakov

The experimental data reported by Professor Kiperman
showing the catalytic properties of nickel cleaned in a common and a superhigh vacuum system seem to be very rarely
observed. In our experiments while studying many catalytic
systems such an effect has been never observed. Thus, A.V.
Sklyarov, while studying CO oxidation on Pt at 10^{-7} torr,
showed that the activity of a catalyst from experiment to
experiment decreased and tended toward its stationary value.
While carrying out experiments in a superhigh vacuum system
such an effect is not observed, and the stationary activity was
set in the very first experiment.

The change of the character of kinetic relationships
resulting from insufficient cleaning of a catalyst was also
observed.

Special properties of very clean metals are supported
by works of many other authors. Thus, Robertson showed
that carefully cleaned metals possess very high activity and
perform catalytic decomposition of saturated hydrocarbons at
a room temperature. The specificity of the adsorption
properties of very clean metals is well known.

The analogy of the mechanism of CO oxidation by oxygen
proposed by us with the mechanism found by Langmuir 50
years ago noted by M.I. Temkin seems to be natural, as
Langmuir's experiments were performed under very clean
conditions which cannot be achieved in most present catalytic
experiments. It is known that the methods of preparing and
carrying out experiments developed by Langmuir form the
basis of present superhigh vacuum installations.

It should be, however, noted that the observed reaction
rates at the same value of a catalyst surface were two to
three orders higher than those in Langmuir's experiments.
Besides, some details of the reaction mechanism in our
work are different, and we were able to obtain a full catalytic
equation of the process which described all our experimental
results well, which Langmuir did not do.

Application of Computers to Kinetic Studies

V. B. Skomorokhov, M. G. Slin'ko and V. I. Timoshenko

Institute of Catalysis, Siberian Branch, USSR Academy of Sciences, Novosibirsk

INTRODUCTION

Computers have already been used for chemical process simulation for more than 12 years. Lately, however, computers have also been used actively in chemical system studies to obtain mathematical models.

At present mathematical simulation methods are fairly well developed. Their practical application, however, is restrained in many cases because of the absence of reliable kinetic models. The construction of kinetic models of complex reactions is a difficult and time-consuming problem, not only experimentally, but also in the computing sense. These circumstances have resulted in the appearance of a new trend in computer applications in chemical studies, consisting of the coupling of the computer and an experimental installation, which provides for the solution of all experimental computing problems in real time and gives an opportunity to use the best methods of work planning and experimental data processing.

In application to kinetic studies, the software of the computer and the experimental installation must provide for:

a) information exchange between the installation and computer, i. e. reception, pre-treatment, accumulation and delivery of results;

b) successive experiment planning for the purpose of choosing one of the competitive hypotheses for the kinetic model and to make its parameters more exact. This planning is carried out by the determination of conditions for each successive experiment which give the maximum information gain about the subject under study. From the viewpoint of mathematics, this problem consists in calculation of the information measure (the information entropy, Coolback's information measure, Fischer's information matrix determinant etc.) and finding the experimental conditions giving its optimal value;

149

c) calculation of the kinetic model parameters from the experimental data by means of minimization of a function acting as a criterion of the coincidence between calculated and experimental rate values;

d) calculation of the parameters of differential and transcendental equation systems on the basis of solutions known for some separate points;

e) kinetic equation analysis to determine the number of independent parameters, the number of stationary regimes etc.

All these programmes must act in the real time-scale.

Elaboration and improvement of such software is a complex and time-consuming problem. The labor expenditure it requires is comparable with that needed for the whole installation; including the computer. This problem cannot be solved without the participation of skilled mathematicians since it implies serious computing difficulties. However these labour expenditures occur only once since on transition from one process study to another it is only necessary to change a few relatively simple nonstandard blocks in the program reflecting the particular reaction being studied.

It should not be thought that application of the computer eliminates the proficiency, experience and intuition of skilled investigators. On the contrary, by releasing him from the routine work of measurement and calculation, the computer greatly expands his creative potential. Computer application requires from the investigator greater clarity in the formulation of goal and the ability to understand the significance of the results obtained and to plan the subsequent experiments.

This paper deals with only two of the great number of possible computer applications in kinetic studies:

1) some experimental problems of reaction rate measurement will be considered using the example of kinetic study of a single-staged dehydrogenation process.

2) the problems connected with calculation of kinetic model parameters will be discussed.

Reaction Rate Measurement

The determination of reaction mixture composition is the most tedious and labour consuming part of the kinetic study of complex chemical processes. Results obtained by conventional physico-chemical methods of analysis are

usually in a form inconvenient for subsequent treatment.
Thus, for example, information on multi-component
mixture composition derived from chromatographic and
mass-spectrometric analysis is coded in a set of peaks,
and its extraction requires measurement of the parameters
of these peaks and the performance of simple but numerous
and therefore time-consuming calculations. In this
connection there appears a disproportion between the
laborious analysis performance and the ease of results
processing. This disproportion grows rapidly as the degree
of complication of the reaction mixture composition
increases and becomes more marked, since the mixture
composition is not a single characteristic of the reaction
under study. As a consequence it is necessary to calculate
the rates of chemical transformations of reactants, to check
the material balances and to average the results of parallel
measurements. Computer applications for automation of
these calculations do not eliminate the above disproportion
completely because the labor-consuming non-automated
step of data preparation for input into the computer remains.
 Direct coupling of the laboratory installation to a
computer is the most effective way of experimental data
processing, enabling us to automate collection, storage and
processing of experimental data completely. The above
conclusions may be illustrated by the rate measurement of
one-step butane dehydrogenation to divinyl,[1] which is a
complex five-route process taking place on a catalyst of
variable activity. Rate measurements were carried out by
a gradientless method in a low-inertia reactor [2] so that
changes of catalyst activity could be followed during the
process. Reaction mixture analysis was accomplished by
means of three coupled chromatographs using two samples.
The duration of each analytical determination was 4 min.
The experimental installation was coupled with a JRA-5
computer (see Fig. 1). Information on reaction mixture
composition was supplied to a commutator as dc potential
proportional to the chromatograph detector signal. The
commutator was connected with an analog-to-digital
converter in a definite pre-programmed sequence. The
converter generated a number which was proportional to
the signal and in machine-coded form this was supplied to
the control device of the computer. Control of the
commutator was performed by the computer via a digital -
to-analog converter. Chromatographic peak indentification

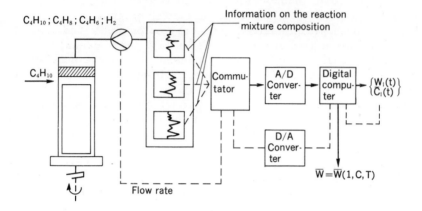

Fig. 1.   Block diagram of experimental installation-computer connection.

was based on the output time counted off from the first peak appearance.

The computer processed the input information directly during the experiment. The processing program provided for chromatographic peak integration, their ascription to the proper substances, calculation of reactant concentrations and conversion rates, verification of the material balances for carbon, hydrogen and argon, and accumulation of the results obtained in a computer memory. Coefficients of polynomials relating reactant concentrations and conversion rates with the catalyst action period were calculated after the experiments were over.

Table I illustrates the correlation of the time needed for carrying out and processing the results of a single experiment manually, by means of a computer, and for a computer coupled with the installation. It shows the high efficiency of such coupling.

The computer-installation coupling imposes certain requirements on both units. The most important is the reliability of the whole set. Its operation will be effective provided that all elements and assemblies of the experimental installation, computer and communication channels are highly reliable. All these elements are equally important as they are continuously interacting parts of a single system,

Table I

| | Mode of Treatment | | |
Stages	Manual	Computer	Computer on line
Experiment as a whole made up of:	40	40	40
Catalyst training	20	20	20
Experiment itself	20	20	20
Results processing made up of:	360	120	1
Peak measurement	60	60	during the experiment
Initial data punching	-	60	-
Calculations	300	1	1
Processing duration Experiment duration	9	3	0.02

and a fault in any one of them results in failure of the whole set. An obligatory demand on the experimental installation is equipment with sonsors permitting acquisition of the necessary information in the form of electric signals. Some types of sensors such as photoelectronic multipliers can generate a machine-coded number directly, but most of them generate analog signals in the form of dc potential. Typical sensors of the latter type are chromatographic detectors, gas and liquid flow sensors based on heat conductivity measurements etc. Some measuring instruments utilize electric current parameters. Thus piezoelectric pressure sensors generate alternating current the frequency of which is proportional to the pressure measured, as a rule. All sensors of this kind are usually equipped with electronic circuits converting the current parameters to voltage.

This makes it possible to use standard analog-to-digital converters as universal information converters. Their high accuracy (fraction percent) and low inertia (conversion time is up to several μsec.) permit them to be used for

automation of the processing of results for measurements
performed by any physico-chemical analysis method.

The information flow density greatly affects the
efficiency of coupling. Slow, time-extended information
receipt results in computer time wastage. The experiments
should therefore be based on high-speed methods of analysis.
From this point of view, continuous concentration measurements are most preferable, e. g., by means of an optical-
acoustic method based on heat conductivity or magnetic
susceptibility measurements etc. The use of capillary
columns and ionization detectors in chromatography, and
electronic but not magnetic mass-spectrum scanning etc.
is desirable.

The structure of the experimental installation must
provide for the possibility of this synchronization with the
computer. This requirement is particularly important when
investigating transient processes and while using
chromatographic methods of analysis, as in this case the
peak identification depends on the output-time.

The computer must meet the requirements of speed of
action and operational memory capacity, which are mainly
determined by mathematical complexity arising during
processing of experimental results, and by accumulated
information volume. If the computer is coupled with several
installations simultaneously it should necessarily have a
system making it possible to select an operation to be given
priority when operating in a time-sharing regime.

Control computers are the most convenient for coupling.
Their assembly includes the computer itself and also analog-
to-digital and digital-to-analog converters and a commutator
for interogating sensors in a predetermined sequence. The
external numerical data input and output blocks placed in the
immediate vicinity to the installation give the experimenter
maximum comfort in his work.

A medium power control computer "DNEPR-2, "
consisting of control ("DNEPR-22) and computing ("DNEPR-
21") complexes, is used for coupling with the experimental
installation in the Institute of Catalysis at the present time.
This machine provides for solving the above problems (see
"Introduction") in real time, including installation operation
control, though on this step the researcher should contribute
his share to the work. Figure 2 shows the general scheme
of machine coupling with the installation for isoamylene
dehydrogenation kinetic study.

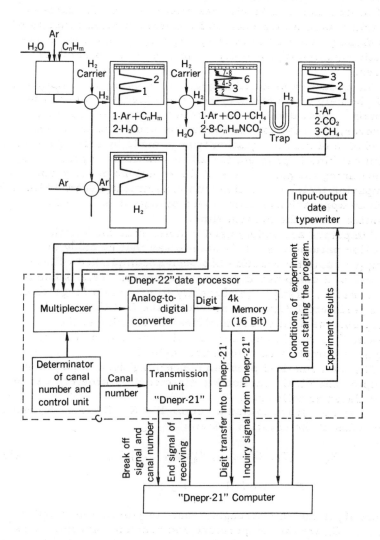

Fig. 2. General scheme of machine coupling with the installation for isoamylene dehydrogenation kinetic study.

Experimental Data Processing

Construction of a kinetic reaction model on the basis of experimental data for chemical conversion rates consists of two closely related and overlapping stages: the choice of a kinetic equation type and determination of the parameter values.

The deduction of kinetic equations is easily accomplished with the help of a stationary reactions theory developed by Temkin[3] provided that the reaction mechanism is known. However, the reaction mechanism is usually unknown and the choice of kinetic equation type is made by solving a direct chemical kinetic problem for all mechanisms suggested, by determining the parameters of the equations deduced and comparing their correlation with experimental data. Although such an approach has many shortcomings it is the only possible one at the present time.

Computers are widely used to determine the kinetic equation parameters. It is safe to say that no study of a complex catalytic reaction has been accomplished without the use of a computer in the last ten years. This is due to two reasons; firstly, the necessity to process a large volume of experimental data, and secondly, the fact that kinetic equations, as a rule, are nonlinear in relation to the parameters sought and therefore determination of these requires complicated calculation methods.

The mathematical problem concerning determination of complex chemical reaction kinetic model parameters can be formulated as follows.

It is necessary to find a vector $K = K_1, K_2, \ldots, K_n$ that is a solution of the equation system:

$$R_{ij} = f_i (T_j, C_{ij}, \ldots, C_{qj}; k_1, \ldots, k_n) \tag{1}$$

$$i = 1_1 P \qquad j = 1_1 N ; \qquad i \cdot j > n$$

where R_{ij} is the rate through i-route in j-experiment and T_j, $C_{1j}, \ldots, C_{qi}$ are the temperature and reactant concentrations in j-experiment, respectively; $k_i, k_2, \ldots, k_n$ are the desired kinetic equations parameters; N is the number of experiments; P is the number of stoichiometrically independent routes; and f_i is a function given.

This is a typical inverse problem - to recover the equation coefficients on the basis of the solution, given for some points.

If the values R_{ij} and C_{ij} were known exactly, the system (1) would be completely compatible with the physical problem. However, this compatibility is broken due to the inevitable experimental errors.

It is therefore possible to find only an approximate solution of this system by minimizing the vector norm of divergences calculated on the base of the system (1) R_{ij} from the experimental values.

It is obvious that the solutions obtained by use of various norms can differ from each other.

If a maximum verisimilitude criterion is taken as a basis we have to find the probability density maximum of joint distribution of all errors. In the case of normal divergence distribution the maximum verisimilitude estimation is achieved if the divergence vector norm is a weighted sum of divergence squares calculated from the measured values.

$$S = \sum_{j}^{N} \sum_{i}^{P} (R_{ij} - \hat{R}_{ij})^2 \tag{2}$$

The sign $\wedge$ is the value calculated from equation (1).

In the case of exponential error distribution, the verisimilitude function is maximum when the other norm is taken, namely, when the weighted sum of the absolute divergence values is minimum.

$$S = \sum_{j}^{N} \sum_{i}^{P} /R_{ij} - \hat{R}_{ij} / \tag{3}$$

In accordcance with the maximum verisimilitude criterion each error distribution correlates with the norm of the divergence vecor. In this case experimental information on chemical conversion rates is utilized with the greatest efficiency.

Thus, selection of a rational method for reaction rate constants determination demands preliminary analysis of the divergence distribution function. In practice an iteration procedure is reasonable, i.e. estimation by one of the methods above and then verification of the divergence distribution function.

The least-squares method is widely used due to the fact that for the linear problem (1) there exist well-developed computing schemes which have also been generalized for nonlinear cases. Different modifications of a gradient method, a method of ravine and a nonlinear estimation method are used for minimization most extensively.

However, experience of many years application of the least-
squares method shows that it does not always give satisfactory
result. A.N. Tikhonov[4] has proved that even for linear
problems this method is incorrect according to Adamar, i.e.,
any small errors in R_{ij} and C_{ij} determination may result in
most significant variations in the solution. For non-linear
problems there is also a possibility of several solutions due
to S-function polyextremality. This results in the repeatedly observed ambiguity of kinetic equation parameters and in
obtaining physically absurd values. In addition, the experimental error is not embodied only in R_{ij} and is not distributed
normally in most cases.

The determination of the sum minimum of the absolute
error values for multi-dimensional functions involves great
calculative difficulties in the case of complex reactions.
Before the creation of linear programming methods and
computers there were no convenient algorithms to solve the
problem. Therefore this estimation method was of little
use.

A norm quite different in principle is the minimum of
maximum divergence proposed by P.L. Chebyshev,

$$\lambda = \min_{k} \max_{i.j} /R_{ij} - \widehat{R}_{ij}/ \tag{4}$$

providing the best uniform approximation of the system of
R_{ij} experimental values whereas the least-squares method
and the least absolute divergence methods give their best
approximation on the average.

Chebyshev introduced this norm instead of the divergence
squares sum minimum more than 100 years ago while creating mechanism theory. His objective was to find the maximum
divergence values at which a mechanical system keeps its
physical properties.

Chebyshev's method was rarely used in practice because
of the complexity of the calculational algorithms. With the
appearance and development of computers and linear programming methods these difficulties no longer arise. It has
been shown (see, for example, ref. 6) that error levelling
according to Chebyshev is equivalent to the linear and convex
programming problems. For the system (1) they are
formulated as follows: to find the minimum value with the
limitations

$$/R_{ij} - f_i (T_j, C_{ij}, \ldots, C_{gj}; k_1, \ldots, k_n)/ \leq \lambda \qquad (5)$$

The main advantage of Chebyshev's method is a conservation of obtained solutions in the physical sense for small variations of initial experimental data, while the least-squares method is highly susceptible to these because of its incorrectness according to Adamar.

The second advantage of Chebyshev's method is its small sensitivity to error distribution law. This is a consequence of Chebyshev's norm having a non-statistic nature. It should be noted that this advantage is very important because the normal error distribution required for correct application of the least-squares method is not observed in kinetic studies in most cases, as already mentioned above (large divergences are absent).

Chebyshev's method also eliminates the second limitation of the least-squares method, i.e., the requirement of the absence of independent variable (concentration and temperature) errors.

This follows from the fact that there is no difference in principle between all measured values with regard to inequalities (4). At the same time, in statistical processing of data in which this requirement is not realized, it is necessary to pass from regressive to confluent analysis which is far more complex and insufficiently developed. It should be noted that solution of Chebyshev's error levelling problem by the linear programming method makes it possible to impose limitations of the type.

$$\alpha_i \leq k_i \leq B_i \qquad (6)$$

on a solution area by means of inclusion of all of these in the inequality system (4).

Solution of dual problems, according to academician Kantorovitch, permits us to determine system sensitivity to experimental errors simultaneously with the parameter determination.[8]

The main shortcoming of Chebyshev's method is the great solution sensitivity to measurement errors. It is possible, however, to construct an iterative process omiting experimental data with great errors at each step and to investigate the solution convergence and sensitivity.

Results of processing of M.N. Andrushkevich's[7]ex-

perimental data on the kinetics of n-butylene oxidative
dehydrogenation on Cr-Ca-Ni-P catalyst by a flow-circulation
method are given as an example of the application of
Chebyshev's error levelling method. Conversion rates of
n-butylenes (W_1), divinyl (W_2), oxygen (W_3), hydrogen (W_4),
carbon monoxide (W_5), and carbon dioxide (W_6) were measured experimentally.

The problem was to calculate the rates through the
following routes:

$$C_4H_8 + 0.5\ O_2 = C_4H_6 + H_2O \tag{7}$$
$$C_4H_8 + 5.5\ O_2 = 4CO_2 + 3H_2O$$
$$C_4H_8 + 3.5\ O_2 = 4CO + 3H_2O$$
$$C_4H_6 + 4H_2 = 4CO + 7H_2$$

To the system (1) corresponded the system:

$$-R_1 = W_1 \tag{8}$$
$$R_1 - R_2 - R_3 - R_4 = W_2; \qquad FR_4 = W_4$$
$$-0.5R_1 - 5.5R_2 - 3.5R_3 = W_3; \qquad 4R_3 + 4R_4 = W_5$$

From physical considerations the limitations (5) were
formulated in the form $0 \leq R_1 \leq 1$, $2\ W_j$. Calculations were
performed by the computer M-20 in accordance with a program elaborated in the Institute of Mathematics of the
Siberian Branch of the USSR Academy of Sciences.

Table II gives the calculation results for two experiments
(according to Chebyshev). The results of calculation by the
least-squares method (column l.s.m.) and by key-substance
method are also given. In the latter case indexes of value W_j
chosen as key ones are given in the corresponding column
heading.

Table II shows that with use of a least-squares method,
negative rates appear though all of the routes are irreversible.
The table also shows that selecting various substances as key
ones can give rather great differences for rates through the
routes. Finally, using Chebyshev's error levlling method,
we obtain rate-through-route values in agreement with the
problem in a physical sense.

Table III lists rate-through-route values calculated
according to Chebyshev and by the least-squares method for
two parallel experiments in which differences in the conversion rates did not exceed 10%. It follows from Table III that
Chebyshev's method is really stable in relation to small

measurement fluctuations, while the solutions obtained by the least-squares method can differ from each other in sign and by more than two orders of magnitude at sufficiently high accuracy (under modern conditions) of chemical conversion rate measurement.

Table II.

No. of experiment	mmole g. hr	l. s. m.	mmole/g.hr					Chebyshev	
			1.2 3.4	1.2 3.5	2.3 5.6	2.3 4.5	1.3 5.6	1.2 4.5	
1	- 9.8	9.3	9.8	9.8	6.3	7.0	9.8	9.8	10.1
	6.6	2.3	4.5	2.1	2.4	2.5	2.4	2.1	2.5
	-23.9	1.6	-1.6	2.1	2.1	1.8	1.7	0.4	1.4
	1.0	-0.8	0.3	-0.9	0	0.7	-0.6	0.7	0.1
	4.6								
	9.5~								
2	-10.4	11.9	10.4	10.4	9.5	7.0	10.4	10.4	9.3
	7.5	0.8	-3.3	1.8	0.9	0.7	0.9	2.1	0.6
	- 7.8	-0.8	5.8	3.6	-0.5	0.7	-0.7	0.7	0.3
	3.0	4.3	0.4	4.7	1.6	0.4	1.8	0.4	0.6
	4.5								
	3.5								

Table III.

Substance conversion rates			Rates through the routes				
Designation	Experiment		l. s. m.			Chebyshev	
	I	II		I	II	I	II
$-W_1$	15.1	15.8	R_1	15.93	17.26	15.12	15.90
W_2	16.6	18.1	R_2	0.28	0.12	1.02	1.20
$-W_3$	10.1	10.4	R_3	0.35	0.34	0.50	0.50
W_4	4.1	3.9					
W_5	2.4	2.2	R_4	0.00074	-0.051	0.4	0.33
W_6	2.5	2.4					

Improved Steady-Rate Treatment of a Chemical Reaction with Regard to Selectivity of Complex Catalysis

K. MIYAHARA

Hokkaido University, Sapporo

INTRODUCTION

A number of methods have been proposed for the mechanistic analysis of steady chemical reactions based on the assumption of the kinetic mass-action law for the rate of each constituent step as well as for the rates of overall reactions.

The kinetic mass-action law assumed in those investigations may be valid and useful in the treatment of most homogeneous reactions since the rate constants are unique with respect to the physical nature of the constituent steps when the concentrations of individual intermediates are known. However, this is not the case in heterogeneous catalysis which involves reactions among adsorbed entities, the concentrations of which during catalysis are usually difficult to determine and which do not obey the simple kinetic mass-action law.

It is thus desirable to improve the method in order to make it applicable to the latter reactions as well. This problem has already been essentially solved by Horiuti[1-3] in the case of reactions over a single pathway (cf. next section) and has been applied successfully to some types of heterogeneous catalysis, e.g., hydrogen electrode processes, ammonia synthesis and the hydrogenation of ethylene. In more complex reactions, however, one has to consider the case of two or more reaction routes and the differences among the rates of formation of these products, i.e., the selectivity. Selectivily is usually attributed to differing amounts and stabilities of intermediates, each of which is supposed to be characteristic of one of the overall reactions proceeding simultaneously. However, such a view of the reaction mechanism is only intuitive, and even useless when the steps involving these intermediates are not rate-determining in the respective reactions.

The principal purpose of the present report is to develop, without any assumptions except steady state conditions of catalysis, a method for the mechanistic analysis of complex catalysis, the constituent steps of which are known. Before going into this application, the basic method will first be described by applying it to some simple cases, i.e., the ammonia synthesis reaction, the steps discussed by Christiansen[4] and the hydrogenation of ethylene catalyzed by nickel.[1,2,5]

STEADY-STATE CONDITIONS FOR A SET OF STEPS AND THE NUMBER OF POSSIBLE OVERALL REACTIONS LINEARLY INDEPENDENT OF EACH OTHER

The steady state of a set of steps is exemplified here by the ammonia synthesis reaction, which is assumed to occur via the following

$$
\begin{aligned}
s=1: \quad & N_2 \rightleftarrows 2N(a), \\
2: \quad & H_2 \rightleftarrows 2H(a), \\
3: \quad & N(a) + 3H(a) \rightleftharpoons NH_3,
\end{aligned}
\tag{1}
$$

where (a) denotes the adsorbed state. This set of steps might be quite different from the real one, but it is sufficient to exemplify the present treatment of steady state conditions. Denoting the intermediates $N(a)$ and $H(a)$ by X_1 and X_2, their concentrations by $[X_1]$ and $[X_2]$ and the net rates of the above steps by v_1, v_2 and v_3, respectively, we have as the steady state condition for the set of steps (1)[6]*

$$
\begin{aligned}
d[X_1]/dt &= 2v_1 \quad\quad - v_3 = 0 \\
d[X_2]/dt &= \quad\quad 2v_2 - 3v_3 = 0
\end{aligned}
\tag{2}
$$

* Such simultaneous equations as Eqs. (2) have been given in many texts, e.g., ref. 6. However, most are concerned with the individual unidirectional rates of constituent steps, but not with their net rates, and treat these equations solely as laborsaving devices, missing such relations as described in the present report.

which is summarized in the matrix equation as

$$\left\{ d[X_i]/dt \right\} = (a_{is}) \left\{ v_s \right\} = 0 \tag{3}$$

where

$$(a_{is}) = \begin{pmatrix} 2 & 0 & -1 \\ 0 & 2 & -3 \end{pmatrix} \quad \text{and} \quad v_s = \begin{pmatrix} v_1 \\ v_2 \\ v_3 \end{pmatrix} \tag{4}$$

We have a single, particular solution of Eq.(3) with respect to $\{v_s\}$, hence the general solution is given as

$$\left\{ v_1, \ v_2, \ v_3 \right\} = V\left\{ 1, \ 3, \ 2 \right\}, \tag{5}$$

where V is an arbitrary constant. The overall reaction specified by the solution (5) is derived by summing the steps (1), each multiplied by its so-called stoichiometric number,[3] ν_s, i.e., 1, 3, and 2, respectively, as follows

$$N_2 + 3H_2 = 2NH_3. \tag{6}$$

The quantity V in Eq.(5) is just the steady rate of this overall reaction.

In a general case of n steps involving m intermediates, it can be shown algebraically that the number of particular solutions of Eq.(3) with regard to $\{v_s\}$ is equal to n - q (=p), where q is the rank of (a_{is}) (i = 1, 2, ..., m, s = 1, 2, ..., n) and p, referred to by Horiuti[3] as the number of independent routes, represents the number of overall reactions which can take place simultaneously in the steady state, linearly independent of each other. We have then as a general solution of Eq.(3)

$$\left\{ v_s \right\} = \sum_p v_p \{\nu_s\}_p, \tag{7*}$$

* The same equation has been given by Temkin,[7] who started his discussion from the steady state condition of the p-th overall reaction,

$$\sum_s \nu_s^{(p)} a_{is} = 0$$

instead of Eq. (2), where a_{is} is the stoichiometric coefficient of X_i in s-th step and $\nu_s^{(p)}$ its stoichiometric number.

where $\{\nu_s\}_p$ is the p-th particular solution of Eq. (3) and V_p is the steady rate of the overall reaction characterized by $\{\nu_s\}_p$.

IMPROVED EXPRESSIONS FOR THE FORWARD AND BACKWARD UNIDIRECTIONAL RATES OF A STEP

We can now express the forward and backward unidirectional rates of any particular step in a general way as a product of two characteristic functions. One is a standard value, usually the upper limit, of the unidirectional rate of the step, defined under the hypothetical condition that all steps preceding or following the step are in equilibrium. The other is the activity of the initial and final system of the step relative to the standard value, also defined under the same hypothetical condition. By the use of these functions we can evaluate the relative rates and affinities of each constituent step, the activities of individual intermediates etc. under the experimental conditions employed, without the use of assumptions currently adopted, as mentioned in the introduction.

We will start our discussion with the statistical mechanical expression[3] for the forward and backward unidirectional rates of the s-th step, respectively

$$v_{+s} = (kT/h) \, a^{I(s)} / a^{\neq(s)} \qquad (8.f)$$

and

$$v_{-s} = (kT/h) \, a^{F(s)} / a^{\neq(s)} \qquad (8.b)$$

where k, T and h have their usual significance, and $a^{I(s)}$, $a^{F(s)}$ and $a^{\neq(s)}$ represent the absolute activity[8] of the initial, the final and the critical system, $I(s)$, $F(s)$ and $\neq(s)$ respectively, of the step s, evaluated by taking into account their concentrations and the interactions between them and other adsorbed entities, if present. The absolute activities, $a^{I(s)}$ etc., are now related to the chemical potentials of $I(s)$ etc. by

$$\mu^{I(s)} = RT \ln a^{I(s)} \qquad (9)$$

etc., and hence the terms $a^{I(s)}/a^{\neq(s)}$ and $a^{F(s)}/a^{\neq(s)}$ in Eq. (8) are expressed as $\exp(-\Delta G^{\neq}_{+s}/RT)$ and $\exp(-\Delta G^{\neq}_{-s}/RT)$, respectively, in terms of the activation free energy, $-\Delta G^{\neq}_{+s}$ or $-\Delta G^{\neq}_{-s}$,

similar to the expressions for the rate-constants given by
Glasstone et al.   Examples of particular expressions for
the absolute activity are given later for hydrogen and
ethylene molecules assumed to behave as an ideal gas.
We can rewrite Eqs. (8.f) and (8.b) as

$$v_{+s} = u_s f^{I(s)} \text{ and } v_{-s} = u_s f^{F(s)} \qquad (10.v)$$

where

$$u_s = (kT/h) a_e^{I(s)} /a^{\neq(s)}, \qquad (10.u)$$

$$f^{I(s)} = a^{I(s)} /a_e^{I(s)}, \quad f^{F(s)} = a^{F(s)} /a_e^{I(s)} \qquad (10.f)$$

and $a_e^{I(s)}$ is a standard value of $a^{I(s)}$, which would be realized
if every step but s was in equilibrium. Since the quantity
$a_e^{I(s)}$ can be given as the product of the a-functions of reactants
and/or products of the overall reaction, from which I(s) is
formed by equilibrium steps, it is uniquely determined by the
experimental conditions employed. The quantity u_s thus defined by Eq.(10.u) is the upper limit of the forward, unidirectional rate, v_{+s}, and hence is a parameter characteristic of step
s uniquely determined by the experimental conditions. Note
that all the quantities given by Eqs.(10) are positive
and $f^{I(s)}$ and $f^{F(s)}$ are usually different from unity, since not
every step is in equilibrium. We see from Eqs.(9) and (10.f)
that

$$RT \ln(f^{I(s)} /f^{F(s)}) = \mu^{I(s)} - \mu^{F(s)} = -\Delta G_s, \qquad (10.G)$$

which is the affinity of step s.

APPLICATION TO THE SET OF STEPS DISCUSSED BY CHRISTIANSEN

Reaction over a Single Route
    The method described above is applied first to the "open
linear sequence of steps" of Christiansen,[4] which is a simple
one such as

$$A_s + X_{s-1} \underset{\longleftarrow}{\overset{\longrightarrow}{}} X_s + B_s,$$

$$(s=1, 2, \ldots, n; X_0 \text{ and } X_n \text{ are absent})$$

where A_s and B_s are the stable reactant and product of step s, respectively, and X_s is the intermediate formed thereby.

The steady state of this set of steps is defined by Eq.(3), where $i = 1, 2, \ldots, n-1$ and

$$(a_{is}) = \begin{pmatrix} 1 & -1 & 0 & \ldots & 0 & 0 & 0 \\ 0 & 1 & -1 & .. & 0 & 0 & 0 \\ & & \cdots\cdots\cdots\cdots \\ 0 & 0 & 0 & .. & 1 & -1 & 0 \\ 0 & 0 & 0 & .. & 0 & 1 & -1 \end{pmatrix}$$

The rank of (a_{is}) is $n-1$ and accordingly we have $p = n - (n-1)$, so that there is a single particular solution of Eq.(3)

$$\{ v_s \} = \{ 1, 1, \ldots, 1 \}$$

which results in the overall reaction

$$\sum_{s=1}^{n} A_s = \sum_{s=1}^{n} B_s \tag{11.R}$$

The steady rate V of this overall reaction is then expressed by reference to Eqs.(10) as

$$\left. \begin{aligned} V &= v_{+s} - v_{-s} \qquad (s = 1, 2, \ldots, n) \\ &= u_1(1 - f_1) = u_2(f_1 - f_2) = \ldots = u_s(f_{s-1} = f_s) \\ &= \ldots = u_n(f_{n-1} - f_n) \end{aligned} \right\} \tag{11.V}$$

where u's are defined similarly to Eq.(10.u) and f's are

$$\left. \begin{aligned} f^{I(s)} &= a^{I(1)} / a_e^{I(1)} = a^{A_1}/a^{A_1} = 1 \\ f_s &= f^{I(s)} = f^{F(s-1)} \\ f_n &= f^{I(n)} = \prod_{s=1}^{n} (a^{B_s}/a^{A_s}) \end{aligned} \right\} \tag{11.f}$$

We see from Eqs. (11.V) that all the f's, which are positive according to their definitions, are smaller than unity as long as $V > 0$. We see, further, from Eqs.(10.G), (11.V) and the last of Eqs.(11.f) that

$$-\sum_{s=1}^{n} \Delta G_s = RT \ln \left[\frac{1}{f_1} \prod_{s=2}^{n} (f_{s-1}/f_s) \right] = RT \ln f_n = -\Delta G \tag{11.G}$$

where $-\Delta G$ is the affinity of the overall reaction (11.R).

Eliminating f_i' s (i=1, 2, ...,S-1) from Eqs.(11.V), we have

$$(1 - f_s)/V = \sum_{s=1}^{s} 1/u_s \quad \text{or} \quad f_s = 1 - V\sum_{s=1}^{s} 1/u_s, \tag{12.u}$$

which is reduced to

$$1/V = \sum_{s=1}^{n} 1/u_s \tag{12.V}$$

in the case where $s \leqq n$ and f_n is negligibly small as compared with unity.*

Equation (12.V) states that the smallest of the u's is the upper limit of V and further that $V = u_r$ in the special case where $u_r \ll u_s$ ($s \neq r$), that is, where step r determines the steady rate of the overall reaction.

If the j-th step is irreversible, i.e., if f_j is negligible compared with f_{j-1}, we have $f_j \ll 1$ since $f_{j-1} < 1$ for $V > 0$. Thus, Eq.(12.u) is reduced in this case to

$$1/V = \sum_{s=1}^{j} 1/u_s$$

and we see that the steady rate V of the overall reaction is independent of the rates of steps which follow the irrevesible step j.

A further discussion is possible on the relation between $- \Delta G_s$ and u_s. It follows from Eqs.(10.G) and (12.f) that

$$- \Delta G_s = RT \ln[(1 - V\sum_{}^{s-1} 1/u_s)/(1 - V\sum_{}^{s} 1/u_s)], \tag{13}$$

from which we see that u for an equilibrium step s' is far larger than V since $- \Delta G_{s'} \cong 0$. Thus, we can evaluate the relative magnitude of ΔG 1 for a non-equilibrium step 1 from that of u1, according to Eq.(13). For instance, in case of $V/u1 = 1/4$ for steps 1 = 1, 2, 3 and 4, it follows from Eq. (13) that

$$\Delta G_2/\Delta G_1 = 1.05, \qquad \Delta G_3/\Delta G_1 = 1.50, \qquad \Delta G_4/\Delta G_1 \gg 1$$

These results show that in a sequence of non-equilibrium steps with the same value of u, the affinity of a particular

* This condition means that the experimental conditions are quite different from the equilibrium of the overall reaction (11.R) as seen from Eq.(11.G).

step is generally smaller than that of the step that follows, or conversely,, for those with common values of the affinity, the first step has the smallest value of u.

Reaction System with Two Routes; Selectivity

We next examine the second case of Christiansen[4] which constitutes a chain reaction accompanied by chain-initiation and chain-termination steps as below.

$$
\begin{array}{ll}
A_1 \rightleftharpoons X_1 + B_1 & \text{chain-initiation} \\[2mm]
A_s + X_{s-1} \rightleftharpoons X_s + B_s & \\
\quad (s=2,\ 3 \text{ and } 4) & \text{chain reaction} \\
A_5 + X_4 \rightleftharpoons X_1 + B_5 & \\[2mm]
A_6 + X_4 \rightleftharpoons B_6 & \text{chain-termination}
\end{array}
\qquad (14)
$$

We have in this case as the particular value of (a_{is}) from Eq. (3)

$$
(a_{is}) = \begin{pmatrix}
1 & -1 & 0 & 0 & 1 & 0 \\
0 & 1 & -1 & 0 & 0 & 0 \\
0 & 0 & 1 & -1 & 0 & 0 \\
0 & 0 & 0 & 1 & -1 & -1
\end{pmatrix}
$$

and two independent routes, from which the general solution of Eq. (3) is given according to Eq. (7) as

$$
\{v_s\} = V_1\{1,\ 0,\ 0,\ 0,\ -1,\ 1\} + V_2\{0,\ 1,\ 1,\ 1,\ 1,\ 0\}
\tag{15.V}
$$

$$
= \{V_1,\ V_2,\ V_2,\ V_2,\ V_2 - V_1,\ V_1\}
$$

The quantities V_1 and V_2 are the steady rates of the overall reactions

$$
A_1 + B_5 + A_6 = B_1 + A_5 + B_6 \tag{15.R_1}
$$

and

$$
\sum_{s=2}^{5} A_s = \sum_{s=2}^{5} B_s \tag{15.R_2}
$$

respectively. Christiansen[4] named these two reactions a priori as "side" and main reactions, respectively, and obtained rather complicated relationships even when steps 2 and 6 were assumed to be irreversible. In the present method, such assumptions are excluded and it is shown that the identity of the "main" reaction depends upon the relative magnitudes of u's and, accordingly, upon the experimental conditions employed.

We have from reference to Eq.(15.V)

$$V_1 = u_1(1-f_1) = u_6(f_4-f_6) \qquad (16.a)$$

$$SV_1 = u_2(f_1-f_2) = u_3(f_2-f_3) = u_4(f_3-f_4) \qquad (16.b)$$

$$(S-1)V_1 = u_5(f_4-f_1f_5) \qquad (16.c)$$

where

$$S = V_2/V_1 \qquad (16.S)$$

$$u_1 \equiv (kT/h)a^{A_1}/a^{\neq(1)} \qquad (s=2,3,4 \text{ or } 5)$$

$$u_s \equiv (kT/h)a^{A_s} a_{\neq}^{e} s^{-1}/a^{\neq(s)}$$

$$u_6 \equiv (kT/h)a^{A_6}a_e^{X_4}/a^{\neq(6)}$$

$$f_i \equiv a^{X_i}/a_e^{X_i} \quad (i=1, 2, 3 \text{ or } 4)$$

$$f_5 = \prod_{s=2}^{5} (a^{B_s}/a^{A_s}) \qquad (16.f_5)$$

$$f_6 = (a^{B_1}a^{A_5}a^{B_6})/(a^{A_1}a^{B_5}a^{A_6}) \qquad (16.f_6)$$

and

$$a_e^{X_i} = \prod_{s=1}^{i} (a^{B_s}/a^{A_s})$$

It follows from Eqs.(16), by elimination of f_1, f_2, f_3 and f_4 there, that

$$(1-f_6)/V_1 = 1/u_1 + S\sum_{s=2}^{4} (1/u_s) + 1/u_6 \qquad (17.a)$$

and

$$(1-f_5)/V_1 = (1-f_5)/u_1 + S\sum_{s=1}^{5} (1/u_s) - 1/u_5 \qquad (17.b)$$

Subtracting Eq.(17.b) from (17.a) side by side, we have

$$S = u_5[f_5/u_1 + 1/u_6 + (f_6-f_5)/V_1] + 1 \qquad (17.S)$$

Positive or negative values of S respectively represent the

directions of the two overall reactions $(15.R_1)$ and $(15.R_2)$
being the same or opposite to each other, and an absolute
value of S larger than unity means that reaction $(15.R_2)$ is
the "main" one. We see from Eq.(17.S) that the value of S
depends upon the magnitudes of u's of three steps, 1, 5, and
6, and the experimental conditions which determine the
values of f_5 and f_6, as seen from Eqs.$(16.f_5)$ and $(16.f_6)$.
We see, further, from Eq.(17.S) that

$$S \cong (u_5 + u_6)/u_6$$

in a case where $f_5 \cong 0$ and $f_6 \cong 0$, which means according to
Eqs.(16), that the experimental conditions are far from the
equilibrium of both overall reactions $(15.R_1)$ and $(15.R_2)$.

HYDROGENATION OF ETHYLENE CATALYZED BY NICKEL

We will now discuss the hydrogenation of ehtylene
catalyzed by nickel as a typical example of heterogeneous
catalysis via a single route. The present analysis is based
on the set of steps

$$
\begin{array}{l}
C_2H_4 \underset{\longleftarrow}{\overset{1}{\longrightarrow}} C_2H_4(a) \\
H_2 \underset{\longleftarrow}{\overset{2}{\longrightarrow}} \begin{cases} H(a) \\ H(a) \end{cases}
\end{array}
\underset{\longleftarrow}{\overset{3}{\longrightarrow}} C_2H_5(a)
\underset{\longleftarrow}{\overset{4}{\longrightarrow}} C_2H_6
\qquad (18.s)
$$

where H(a), $C_2H_4(a)$ and $C_2H_5(a)$ are hydrogen atom, ethylene
and ethyl radical in their adsorbed state, respectively.
This scheme was deduced from analysis of experimental
results obtained by infra-red absorption[11] and mass-spectrometric analysis of deuteration of light ethylene[12] Application
of the present method to the set of steps (18.s) easily
demonstrates that the possible overall reaction is single,
i.e.,

$$C_2H_4 + H_2 = C_2H_6 \qquad (18.R)$$

It follows in the steady state that

$$
\left.
\begin{array}{l}
V = v_{+s} - v_{-s} \quad (s=1, 2, 3 \text{ and } 4) \\[6pt]
\quad = u_1(1-f_1) = u_2(1-f_2^2) = u_3(f_1 f_2 - f_3) = u_4(f_2 f_3 - f_4)
\end{array}
\right\} (18.V)
$$

where, if we denote the quantities relating to C_2H_4 and H_2 in the gas phase by the subscripts E and H, respectively,

$$u_1 \equiv (kT/h)\, a_E/a^{\neq(1)} , \qquad\qquad u_2 \equiv (kT/h)\, a_H/a^{\neq(2)} \left.\right\}$$
$$u_3 \equiv (kT/h)\, a_E a_H^{\frac{1}{2}}/a^{\neq(3)} , \qquad u_4 \equiv (kT/h)\, a_E a_H/a^{\neq(4)} \left.\right\} (18.u)$$

$$f_1 \equiv a^{C_2H_4(a)}/a_E , \qquad\qquad f_2 \equiv a^{H(a)}/a_H^{\frac{1}{2}} \left.\right\}$$
$$f_3 \equiv a^{C_2H_5(a)}/a_E a_H^{\frac{1}{2}} , \qquad f_4 \equiv a^{C_2H_6}/a_E a_H \left.\right\} (18.f)$$

As easily seen from Eq.(18.f), RT ln f is the affinity, $-\Delta G$, of the overall reaction (18.R) given as

$$RT\, \ln f_4 = RT\, \ln[(P_{C_2H_6}/P_E P_H)/K_p] = -\Delta G$$

were P is the partial pressure. Thus, f_4 is negligibly small compared with unity under usual experimental conditions, due to the exceedingly large value of the equilibrium constant K_p of reaction (18.R).

Quantities a_H and a_E are now given statistical mechanically as

$$a_H^{-1} = F_H/C_H$$
$$= \frac{kT}{1333P_H} \cdot \frac{(2\pi m_H kT)^{\frac{3}{2}}}{h^3} \cdot \frac{4\pi I_H kT}{h^2} \cdot \exp(-\varepsilon_H/RT) \quad (19.H)$$

and

$$a_E^{-1} = F_E/C_E$$
$$= \frac{kT}{1333P_E} \cdot \frac{(2\pi m_E kT)^{\frac{3}{2}}}{h^3} \cdot \frac{2\pi^2(2\pi I_E kT)^{\frac{3}{2}}}{h^3}$$
$$\prod_j [1 - \exp(h)\nu_j/kT)]^{-1} \cdot \exp(-\varepsilon_E/RT) \quad (19.E)$$

where F is the complete partition function of the gas molecule per unit volume, C is the concentration given as 1333P/kT (P in mmHg), m the mass of a molecule, I_H the moment of inertia of H_2 molecule, I_E the geometric mean of the three principal moments of inertia of the C_2H_4 molecule and ε is the molar energy in the ground state. Rewriting the quantities a_H and a_E in the form of $(P_i/Q_i)\exp(\varepsilon_i/RT)$ (i=H_2 or C_2H_4) and comparing this form with Eq.(19.H) and (19.E), we see that Q_i is constant at a given temperature. The functions(u's) given by Eqs.(18.u) are now recast in the form

$$u_1 = \rho (P_E /Q_E) \exp (-E_1^{\ast}/RT) \equiv U_1 P_E$$

$$u_2 = \rho (P_H /Q_H) \exp (-E_2^{\ast}/RT) \equiv U_2 P_H$$

$$u_3 = \rho (P_E /Q_E) (P_H /Q_H)^{1/2} \exp (-E_3^{\ast}/RT) = U_3 P_E P_H^{1/2} \qquad (20)$$

$$u_4 = \rho (P_E /Q_E) (P_H /Q_H) \exp (-E_4^{\ast}/RT) = U_4 P_E P_H$$

where ρ is a constant approximated to be common to all the u's, $E_s^{\ast}$ is the activation energy of step s for its forward direction, and U's are constant at a given temperature as long as $E_s^{\ast}$ are practically independent of P_H and P_E.

It follows from Eq.(18.V), similarly to Eq.(12.V), that

$$(1-f_4)/V = f_2^2 /u_1 + 1/u_2 + f_2/u_3 + 1/u_4 \qquad (21.a)$$

and, further, on reference to Eqs.(20),

$$P_H/V = f_2^2 P_H/P_E U_1 + 1/U_2 + f P_H^{1/2}/P_E U_3 + 1/P_E U_4, \qquad (21.b)$$

where f_4 is neglected compared with unity.

zur Strassen[13] observed the rate of catalyzed hydrogenation of ethylene in the presence of nickel catalyst under a few hundredths mmHg partial pressures of H_2 and C_2H_4 and found that the steady rate V is strictly proportional to P_H at constant temperature and constant P_E throughout his observations from -7° to 125°C. Similar kinetics of the reaction has been observed by many investigators[5]. It follows from this experimental result that P_H/V of Eq.(21.b) at constant P_E must be practically constant independent of the partial pressure P_H. Since terms U_1 etc. on the right-hand side of this equation are constant, f_2 must be either constant or far smaller than unity according to the relation derived from the first equation of (18.V) and u_2 of (20), i.e.,

$$V/P_H = U_2(1-f_2^2).$$

The first and the third terms of Eq.(21.b) must then be negligibly small as compared with the second and the fourth terms, which are both constant, and hence we have

$$1/V = 1/u_2 + 1/u_4. \qquad (22)$$

One of the characteristics of the reaction in question is the presence of an optimum temperature, T_X, below and above which the initial rate of hydrogenation decreases, i.e., the activation heat is positive and negative, respectively. It was found[12]that T_X is ca. 30°C for the deuteration of light ethylene with equimolar deuterium gas over a nickel catalyst at 0.1 mmHg total pressure. Further, light hydrogen in ethylene is never transferred into D_2 gas during deuteration at -45°C and -23°C, but transferred rapidly at temperatures above T_X, yielding isotopically equilibrium mixtures of hydrogen. The deuterium distribution in ethylene is always random, independent of the reaction temperature being below or above T_X. Such a random distribution of deuterium in ehtylene shows, on reference to scheme (18.s), that steps 1 and 3 must always be in equilibrium. Hence it follows that f_1 of Eq.(18.f) must be unity and further that V/u_1 and V/u_3 are negligibly small, in harmony with the conclusion of Eq.(22). The fact that light hydrogen is never transferred into D_2 gas shows that step 2 is rate-determining at temperatures far below T_X but in equilibrium above T_X. Consequently, we see on reference to Eq.(22), that

$$V = \begin{cases} u_2 \text{ at } T \ll T_X \\ u_4 \text{ at } T \gg T_X . \end{cases} \qquad (23)$$

According to the above conclusion, $E_2^\ast$ in the expression for u_2 in Eqs.(20) can be evaluated by observing V far below T_X, identifying V with u_2 in this region; the value is[5]

$$E_2^\ast = 4.7 \text{ kcal/mole} \qquad (24.2)$$

The quantities $E_3^\ast$ and $E_4^\ast$ are determined according to an equation derived from Eq.(18.V) with $f_1 = 1$ and $f_4 = 0$, i.e.,

$$u_2 : u_3 : u_4 = (1-f_2^2)^{-1} : (f_2-f_3)^{-1} : (f_2 f_3)^{-1}. \qquad (24.f)$$

We can now determine f_2 and f_3 from the observed rates of evolution of individual deutero-substituted ethylenes and ethanes at the initial stages of deuteration of light ethylene, tanking steady state conditions with respect to individual deutero-substituted intermediates into account. It was shown[5] according to the experimental results obtained by Turkevich et al[14] on the deuteration of 10 mmHg light ethylene by 20 mmHg D_2 over nickel wire at 90°C, that the ratio mentioned above takes the following values,

$$u_2 : u_3 : u_4 = 1 : 14.7 : 1.84. \qquad (24.u)$$

From Eqs.(20), (24.2) and (24.u), we obtain

$$E_3^* = -12.6 \text{ kcal/mole} \qquad (24.3)$$

and

$$E_4^* = -18.3 \text{ kcal/mole} \qquad (24.4)$$

Introducing the values of E_2^* etc. derived above and those of Q_H and Q_E into Eqs.(20), we have

$$\left.\begin{aligned}
\log_{10} u_1 &= \infty, \\
\log_{10} u_2 &= \log_{10}\rho(P_H/1.057\times10^{10}) - 1027/T, \\
\log_{10} u_3 &= \log_{10}\rho(P_H/1.057\times10^{10})^{1/2}(P_E/3.939\times10^{14}) + 2751/T, \\
\log_{10} u_4 &= \log_{10}\rho (P_H/1.057\times10^{10}) (P_E/3.939\times10^{14}) + 4008/T.
\end{aligned}\right\} \quad (25)$$

Further, according to Eqs.(18.V), we can determine V, f_2 and f_3, and hence v_{+s} and v_{-s} for steps 2, 3 and 4 individually from the rates u_1 etc. and the values $f_1 = 1$ and $f_4 = 0$.

It is thus found[2,5] that the methods described above are effective in presenting a consistent interpretation of experimental facts obtained on hydrogenation as well as related reactions, e.g., equilibration of hydrogen isotopes, $H_2 + D_2 = 2HD$, para-hydrogen conversion, and hydrogen exchange between light ethylene and deuterium including rates of evolution of individual deutero-substituted hydrocarbons during deuteration of light ethylene.

From Eq.(22) we see further that

$$u_2 = u_4 \quad \text{at} \quad T = T_X,$$

hence that

$$E_2^* - E_4^* = 2.3 \text{ RT}_X \log_{10} (Q_E/P_E),$$

on reference to the expressions for u_2 and u_4 given by Eqs.(20). Comparing this equation with our experimental result[15] that

$$1/T_X = (1/5,300)(14.5 - \log_{10}P_E),$$

we have

$$E_2^{\ne} - E_4^{\ne} = 24.2 \text{ kcal/mole} \qquad (26.E)$$

and

$$\log_{10}Q_E = 14.5. \qquad (26.Q)$$

It follows from Eqs.(26.E) and (24.2) that

$$E_4^{\ne} = -19.5 \text{ kcal/mole}$$

which agrees well with the value given in Eq.(24.4). The value in Eq.(26.Q) also agrees extremely well with the theoretical value, $\log_{10}Q_E = 14.596$.

OXIDATION OF ETHYLENE OVER A SILVER CATALYST

The present method will now be applied to the kinetic analysis of oxidation of ethylene, catalyzed by silver. A view currently accepted on the selectivity for ethylene oxide formation is that the adsorbed state of oxygen resulting in ethylene oxide formation differs from that for complete oxidation of ethylene to CO_2 and H_2O, and the selectivity depends upon the relative amounts. This view, however, has not yet been proved experimentally and further kinetic analysis of the mechanism is required.

We will apply the present method of analysis to the time course of ethylene oxide and carbon dioxide production; these are assumed to be produced by the following set of steps,

$$
\begin{aligned}
&s=1: & O_2 &\rightleftharpoons 2O(a), \\
&2: & C_2H_4 + O(a) &\rightleftharpoons C_2H_4O, \\
&3: & C_2H_4 + 2O(a) &\rightleftharpoons 2CH_2O(a), \\
&4: & CH_2O(a) + 2O(a) &\rightleftharpoons CO_2 + H_2O.
\end{aligned} \qquad (27.s)
$$

This set of steps is simplified from that of Twigg[16] where the adsorbed state of oxygen is assumed to be common to the partial and complete oxidations of ethylene, and the steps for the secondary, complete oxidation of ethylene oxide are excluded.

In the same way as before, we have two independent routes, from which the general solution of Eq. (3) is given as

$$\{ v_s \} = V_1\{1,\ 2,\ 0,\ 0\} + V_2\{3,\ 0,\ 1,\ 2\ \} \qquad (27.V)$$

where V_1 and V_2 are, respectively, the steady rates of the overall reactions

$$O_2 + 2C_2H_4 = 2C_2H_4O \quad \text{(partial oxidation)} \qquad (27.R_1)$$

and

$$3O_2 + C_2H_4 = 2CO_2 + 2H_2O \quad \text{(complete oxidation)}. \qquad (27.R_2)$$

Denoting $O(a)$ and $CH_2O(a)$ by X_1 and X_2, respectively, we can now recast Eq. (27.V) in the forms of

$$\left.\begin{aligned}
(S+3)V_2 &= u_1(1-f_1^2) \\[2mm]
2SV_2 &= u_2(f_1-F_1) \\[2mm]
2V_2 &= 2u_3(f_1^2 - f_2^2) = u_4(f_1^2 f_2 - F_2)
\end{aligned}\right\} \qquad (28.V)$$

where

$$S \equiv V_1/V_2 \qquad (28.S)$$

$$u_s \equiv (kT/h)a_e^{I(s)}/a^{\neq(s)} \quad (s=1,\ 2,\ 3\ \text{or}\ 4) \qquad (28.u)$$

$$f_i \equiv a^{X_i}/a_e^{X_i} \quad (i=1\ \text{or}\ 2) \qquad (28.f)$$

$$\left.\begin{aligned}
a_e^{I(1)} &= a^{A_1} \qquad a_e^{I(2)} = a^{A_1}a_e^{X_1} \\[2mm]
a_e^{I(3)} &= a^{A_2}(a_e^{X_1})^2, \quad a_e^{I(4)} = (a_e^{X_1})^2 a_e^{X_2}
\end{aligned}\right\} \qquad (28.I)$$

$$a_e^{X_1} = (a^{A_1})^{1/2}, \qquad a_e^{X_2} = (a^{A_1}a^{A_2})^{1/2} \qquad (28.X)$$

$$F_1 \equiv a^{B_1}/(a^{A_1})^{1/2}\,a^{A_2}, \quad F_2 \equiv a^{B_2}a^{B_3}/(a^{A_1})^{3/2}(a^{A_2})^{1/2} \qquad (28.F)$$

Gaseous O_2, C_2H_4, C_2H_4O, CO_2 and H_2O, are denoted by A_1, A_2, B_1, B_2 and B_3, respectively. We easily see from Eq. (9) that

$$2RT \ln F_1 = 2\mu^{B_1} - (\mu^{A_1} + 2\mu^{A_2})$$

and

$$2RT \ln F_2 = 2\mu^{B_2} + 2\mu^{B_3} - (3\mu^{A_1} + \mu^{A_2})$$

which are the affinities of the overall reactions (27.R_1) and (27.R_2), respectively. Both F_1 and F_2 are negligibly small compared with unity or with f_i, due to the exceedingly large values of the equilibrium constants of these overall reactions.

We have carried out some experiments on this reaction at temperatures near 300°C using a closed reaction apparatus equipped with a circulation pump. Both V_1 and V_2 are found to be proportional to $P_{O_2} P_{C_2H_4}$ in the time course of a reaction run and their activation heats amounted to 11.0 and 18.5 kcal/ mole, respectively, at an early stage of reaction of equimolar mixture of O_2 and C_2H_4. A typical example of the time course of production of individual gaseous components and S at 300°C is shown in Fig. 1.

In Fig. 1, S as evaluated from the smoothed curves is constant during the early stages of the reaction. Hence we may conclude that the secondary oxidation of ethylene oxide is negligible at this stage; this is consistent with the use of the set of steps (27. s) as the basis of the present analysis.[*]

We will discuss first the experimental results on the dependence of V_1 and V_2 upon the partial pressures, P_{A_1} and P_{A_2}, of O_2 and C_2H_4, respectively, and their activation heats as obtained above. It follows from the first two equations of Eqs.(28.V) and (28.S), neglecting F_1 as compared with f_1,, that

$$1/V_1 = (S+3)/Su_1 + 2f_1/u_2 \qquad (29.1)$$

Similarly, neglecting F_2 as compared with $f_1^2 f_2$, we have

$$1/V_2 = (S+3)/u_1 + 1/u_3 + 2f_2/f_1^2 u_4 \qquad (29.2)$$

Analogous to Eqs.(20), we can write u's, with reference to

[*]   To make the present method of analysis applicable to the whole course of the reaction, some additional steps should be taken into account for the complete oxidation of ethylene oxide, as given by Twigg and others.

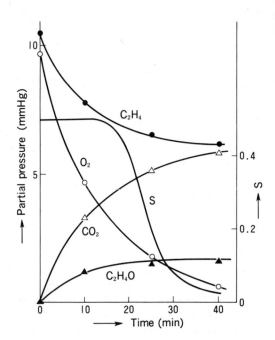

Fig. 1.   The time course of the amounts of individual
gaseous components and the selectivity S ($=V_1/V_2$) during
the oxidation of ethylene by silver catalyst at 300°C.

Eqs. (28. u), (28. I), and (28. X), as

$$u_1 \equiv U_1 P_{A_1}, \qquad\qquad u_2 \equiv U_2 P_{A_1}^{1/2} P_{A_2}$$

$$u_3 \equiv U_3 P_{A_1} P_{A_2}, \qquad\qquad u_4 \equiv U_4 P_{A_1}^{3/2} P_{A_2} \qquad (29. u)$$

The quantities f_1 and f_2 are similarly given as

$$f_1 \equiv a^{X_1}/(U_{A_1} P_{A_1})^{1/2}, \qquad\qquad f_2 \equiv a^{X_2}/(U_{A_1} U_{A_2} P_{A_1} P_{A_2})^{1/2},$$

where $U_{A_1} P_{A_1}$ and $U_{A_2} P_{A_2}$ are the absolute activity of O_2 and
C_2H_4, respectively, as already mentioned in connection

with Eqs. (10) and (20). Consequently, it follows from Eqs. (29. 1) and (29. 2) that

$$1/V_1 = (S+3)/SU_1 P_{A_1} + 2a^{X_1}/U_2 U_{A_1}^{1/2} P_{A_1}^{3/2} P_{A_2} \qquad (30.1)$$

and

$$1/V_2 = (S+3)/U_1 P_{A_1} + 1/U_3 P_{A_1} P_{A_2} + 2U_{A_2}^{1/2} a^{X_2}/U_4 U_{A_1}^{1/2} (a^{X_1})^{1/2} P_{A_1}^{1/2} P_{A_2}^{3/2} \qquad (30.2)$$

The quantities U_s (s=1, 2, 3 and 4), U_{A_1} and U_{A_2} in these equations are constant at a given temperature, whereas the absolute activities a^{X_1} and a^{X_2} may depend upon the partial pressures of the reacting substances. Thus, the observed results that both V_1 and V_2 are proportional to $P_{A_1} P_{A_2}$ show, at least, that the first term in the right-hand side of Eqs. (30.1) or (30.2) is negligibly small compared with the other terms so long as S remains constant. Accordingly, it is concluded that the steady rates V_1 and V_2 are mainly controlled by steps 2 and 3 (or 4), respectively, in agreement with the observation that the activation heats of the overall reactions are different from each other. Further, in Eq. (30.1), the absolute activity a^{X_1} of the intermediate X_1 is concluded to be proportional to $P_{A_1}^{1/2}$, which means that step 1 is in equilibrium, in agreement with the negligibly small value of the first term on the right-hand side of this equation.

In addition to the qualitative conclusions derived above, the relative magnitudes of u's can be evaluated as below on the basis of the observed time courses of the amounts produced of individual gaseous components (shown in Fig. 1). We have from the first equation of (28.V) that

$$f_1 = [1 - (S+3)V_2 /u_3]^{1/2}$$

It follows from the second of Eqs. (28.V) by substituting the expression for f_1 given above and that in (29.u) for u_1 and u_2, that

$$2SV_2 /U_2 P_{A_1}^{1/2} P_{A_2} = [1 - (S+3)V_2 /U_1 P_{A_1}]^{1/2} \qquad (31.1)$$

Similarly, we have from the third and the fourth of Eqs. (28.V) and from (29.u), by eliminating f_1 and f_2,

K. Miyahara

$$V_2 /U_3 P_{A_1}P_{A_2} = 1 - (S+3)V_2 /U_1 P_{A_1}$$

$$- (4V_2^2 /U_4^2 P_{A_1}P_{A_2}) /[1 - (S+3)V_2 /U_1 P_{A_1}]^2 \qquad (31.2)$$

Provided that U's remain constant for small variation of P_{A_1}
etc., we can evaluate U's and accordingly u's by Eqs.(29.u)
from two sets of values of S, P_{A_1} and P_{A_2} observed at two
different reaction times. The results are shown in Table I.
They show that none of the u's is negative and their values
are in conformity with the qualitative conclusions derived
above. These facts demonstrate the effectiveness of the
present method of analysis and of the set of steps (27.s).

The results in Table I show further that U_1 remains nearly
constant, whereas U_2 and U_3 are diminished by ca. 0.46 times,
with a rise in P_{O_2} from 30 to 50 mmHg. By the same approximation used in the foregoing section with respect to
Eqs.(20), U_2 and U_3 are now given as

$$U_2 = (\rho/Q_{A_1}^{1/2}Q_{A_2})\exp(-E_2^{\ddagger} /RT)$$

and

$$U_3 = (\rho/Q_{A_1}Q_{A_2})\exp(-E_3^{\ddagger} /RT)$$

respectively, where ρ is a constant approximated to be
common to all u's, Q_{A_1} and Q_{A_2} are the constants characteristic

Table I.   Calculated Values of U's and u's in the Steady-state
Oxidation of Ethylene at 300°C Catalyzed by Silver
Catalyst.

Initial partial pressure (mmHg)		Mean time of reaction (min)	U_1	U_2	U_3	U_1	U_2	U_3^a
O_2	C_2H_4					(mmHg/min)		
30	10	2	0.24	1.82	9.0	6.6	0.9	2.0
		6		$\times10^{-2}$	$\times10^{-3}$	6.0	0.8	1.6
50	10	2	0.37	0.86	4.0	18.	0.6	0.7
				$\times10^{-2}$	$\times10^{-3}$	17.	0.5	0.7

a   The value of u_4 is always infinitely large since $1/U_4 \cong O$.

of O_2 and C_2H_4 molecules, respectively, and $E_s^{\ddagger}$ is the activation energy of step s for its forward direction. It follows then that

$$0.46 = \exp(-\delta E_s^{\ddagger}/RT) \quad (s=2 \text{ or } 3)$$

where $\delta E_s^{\ddagger}$ is the increment of $E_s^{\ddagger}$ caused by the increase of P_{O_2} from 30 to 50 mmHg, and hence that

$$\delta E_s^{\ddagger} = 0.9 \text{ kcal/mole}$$

This result suggests the presence of a remarkable, repulsive interaction between the critical systems, $\ddagger(2)$ and $\ddagger(3)$, and their surrounding $O(a)$'s, which may increase with rise of P_{O_2} and thus enlarge the values of $E_2^{\ddagger}$ and $E_3^{\ddagger}$. From the same point of view, interaction among $O(a)$ s may be concluded to be absent since the value of U_1 remained nearly constant independent of P_{O_2}.

The activation energies of the steps are estimated on the basis of the temperature dependences of U_2 and U_3 at $P_{O_2} = P_{C_2H_4} = 10$ mmHg as

$$E_2^{\ddagger} = 10.5 \text{ kcal/mole}, \quad E_3^{\ddagger} = 16.5 \text{ kcal/mole}$$

which are close to those observed with respect to the overall reactions $(27.R_1)$ and $(27.R_2)$. The value of $E_1^{\ddagger}$ could not be estimated because of the strongly divergent values of u_1 from run to run in the experiment; this fact probably relates to the observation that the kinetics of the initial rate of the present reactions are not reproducible, strongly depending upon the pretreatment of the catalyst.

In addition to the analysis described above, it is noted that f's for the intermediates can be evaluated by introducing the values of u's into the equations, which are derivable in the same way as Eqs.(28.V) by taking into account some additional steps for the complete oxidation of ethylene oxide. Hence the variations of the activities of intermediates or the affinities of the respective steps during the course of the reaction can essentially be estimated.

The author is much indebted to Mr. Shin-ichi Yokoyama who carried out the experiments and calculations for the oxidation of ethylene catalyzed by silver.

184 K. Miyahara

References

1) J. Horiuti, Bull. Chem. Soc. Japan, 13, 210 (1938).
2) J. Horiuti, Shokubai (Catalyst), 2, 1 (1947);
   J. Res. Inst. Catalysis, Hokkaido Univ. (Sapporo,
   Japan) 6, 250 (1958).
3) J. Horiuti and T. Nakamura, Z. Physik. Chem.
   N. F., 11, 358 (1957); in Advances in Catalysis,
   Vol. 7, p. 1, Academic Press, New York, 1967.
4) J. A. Christiansen, in Advances in Catalysis, Vol. 5,
   p. 311, Academic Press, New York, 1953.
5) J. Horiuti and K. Miyahara, NSRDS-NBS 13 of
   Monograph Series of National Bureau of Standards,
   U. S. A. (1968).
6) S. M. Benson, The Foundations of Chemical Kinetics,
   McGraw-Hill, New York, 1960, p321.
7) M. I. Temkin, Dokl. Academic Nauk, USSR, 152,
   156 (1963).
8) R. H. Fowler and E. A. Guggenheim, Statistical
   Thermodynamics, Cambridge Univ. Press, 1939,
   p. 66.
9) S. Glasstone, K. J. Laidler and H. Eyring, The
   Theory of Rate Processes, McGraw-Hill, New York-
   London, 1941, Eq. (157) on p. 195.
10) e. g., R. M. Noyes, in Progress in Reaction Kinetics
    (G. POTER, ed.), Vol. 2, Chapter 8 (1964).
11) W. A. Pliskin and R. P. Eischens, J. Chem. Phys,
    24, 482 (1956); Advan. Catalysis, 10, 1 (1958);
    Z. Elektrochem., 60, 872 (1956).
12) K. Miyahara, J. Res. Inst. Catalysis, Hokkaido
    Univ. (Sapporo, Japan), 14, 134 (1966).
13) H. zur Strassen, Z. Physik. Chem., A169, 81 (1934).
14) J. Turkevich, F. Bonner, D. O. Schissler and
    P. Irsa, Discuss. Farad. Soc., 8, 352 (1950);
    Colloid. Chem., 55, 1078 (1951).
15) S. Sato and K. Miyahara, J. Res. Inst. Catalysis,
    Hokkaido Univ. (Sapporo, Japan), 13, 10 (1965).
16) G. H. Twigg, Trans. Farad. Soc., 42, 284 (1946).

On the Problem of Classifying Catalytic Reactions

V. A. ROITER and G. I. GOLODETS

L. V. Pisarzhevsky Institute of Physical Chemistry, Academy of Sciences of the
Ukrainian SSR, Kiev

The most important problem in the theory of catalysis is
to predict the rates of catalytic chemical reactions on the
basis of the elementary properties of the catalysts and
reagents. Great difficulties encountered in the quantum
chemical calculations of potential energy surfaces for catalytic processes prevent us from solving this problem at
present.

Inadequate knowledge about the elementary mechanisms
of catalytic reactions is hindering the development of a
quantitative theory.

In developing the modern theory of the reactivity of substances in different noncatalytic reactions, the rules
obtained from the systematization of experimental data in a
deductive manner and their rational classification according
to the type of mechanism are of primary importance. In
developing the chemical structure and kinetics theories as
well as in elucidating reaction mechanisms these rules have
served as the basis for further theoretical generalizations.

It seems natural to go the same way in creating the theory of catalysis, i. e. , first of all to elaborate the detailed
system of the rational classification of catalysts and catalytic reactions and to search for rules connecting chemical
and catalytic properties of substances within the groups of
catalytic systems of the same type. One can hope that
making this classification more precise and interpreting it
in terms of modern conceptions on catalytic mechanisms
will lead to the creation of the general theory of catalytic
behavior prediction. Thus developing (from empirical
generalizations to a quantitative theory) a theory of catalysis
will always be useful in the practice of catalyst selection
and, at the same time, be enriched by this practice.[1]

The necessary prerequisite for a scientific classification
is the systematization of a large body of empirical material
on catalytic properties of substances. With this aim the
Handbook "Catalytic Properties of Substances" has been
compiled in our laboratory,[2] the two first volumes of it

embracing publications up to 1967.*   The data on catalytic
properties in the Handbook have been systematized according to the position of elements forming catalysts in the
Mendeleev Table.   This scheme reflects the fundamental
significance of chemical factor in catalysis.

It is quite clear that to work out the rational classification of all catalytic processes is a very complicated task.
It is natural to begin with the classification of great groups
of catalytic reactions and catalysts and then proceed to
more and more general constructions.

At first, we chose an extensive class of gas phase heterogeneous catalytic reactions involving molecular oxygen.[3,4]
Some of these are of commercial importance.   Numerous
processes of oxidation of different molecules, the decomposition of oxygen-containing compounds with the liberation
of molecular oxygen and oxygen isotopic exchange processes
belong to this class.

All these processes have some common features which
enable us to consider them as a single great class of catalytic reactions.   The participation of molecular oxygen, as
a common reagent, possessing strong electron-acceptor
properties, determines a typical redox behavior of reactions
of this class.   Oxidized molecules usually act as electron
donors.   The electron transition from the molecules to O_2
proceeds via metal or semiconductor catalyst which acti-
vates both reagents on its surface.   The active catalysts of
the reactions considered involve elements which can easily
change their degree of oxidation.   In the formation of
activated complexes at limiting steps in the reactions of this
class, a catalyst-oxygen bond usually breaks or appears.
Therefore, catalytic activity significantly depends upon this
bond strength.[4-9]

However, heterogeneous catalytic reactions involving
O_2 display rather a wide variety of transformations.
Selective oxidation of one molecule is often catalyzed by
essentially different catalysts.   So, the butene-1 oxidation
to methylvinylketone is catalyzed by cuprous oxide; the
oxidative dehydrogenation of the same butene-1 to butadiene
is accelerated by bismuth molybdate; the oxidation of
butene-1 to maleic anhydride is catalyzed by vanadium oxide
catalysts.   The active catalysts of the complete combustion

* Now the third volume including publications of 1968 - 70
is being compiled.

of butene-1 are Pt, Pd, Co_3O_4. These facts suggest that the reactions of the class under consideration proceed via several mechanisms.

To arrive at a rational classification, one must arrange catalytic reactions on the basis of similarity of mechanism, composition and structure of activated complexes. Unfortunately, the detailed mechanisms of heterogeneous catalytic oxidation have not yet been established. Therefore, in our search for a rational classification, we have to use some indirect signs of the similarity of mechanisms.

When the reactions being compared follow the same mechanism, the criteria for optimum catalyst (i.e., the catalyst with the greatest specific activity for the given reactions) should be nearly the same and the similarity of catalytic activity patterns and that of optimum catalysts have to be observed. This fact may serve as the basis for attributing a given group of catalytic reactions to a certain type. This approach enables us to classify not only separate reactions but "reaction-catalyst" systems, i.e., properties of all participants in a catalytic process are taken into account.

A brief form of our proposed classification is given in Table I. Its primary unit is a "reaction-catalyst" system. The combination of such systems involving a given reaction and a number of catalysts of the same type used in this reaction forms a catalytic series. Series of the same type including similar reactions are combined in catalytic types by the similarity of catalytic activity patterns and optimum catalysts. The catalytic types are covered by catalytic subclasses on the basis of the following principle: optimum catalysts of reactions of a given subclass consist of elements or compounds of elements belonging to the same groups of the Mendeleev Periodic System. The combination of these subclasses forms a single class of gas phase heterogeneous catalytic reactions involving O_2.

Let us consider the rational basis of the given classification.

Its first subclass (A) involves relatively simple processes (oxygen isotopic exchange, decomposition with oxygen liberation, complete oxidation) which proceed in a single possible direction or in the most advatageous thermodynamic way. In this case the selectivity problem is practically absent and the problem of the greatest catalytic activity acquires special significance.

Table I   The Classification of Gas-Phase Heterogeneous
Catalytic Reactions Involving Molecular Oxygen.

Catalytic Subclasses	Catalytic Types		Optimum Catalysts
A. The processes greatly accelerated by catalysts involving elements or compounds of elements belonging mostly to the VIII (also to Ib) group of the Mendeleev System (Pt, Pd, Co, Ni, Cu)	A. I.	Isotopic exchange of oxygen-containing molecules with O_2	Pt, Pd, Ni, Co_3O_4, $CuCo_2O_4$
	A. II.	Decomposition of unstable oxygen compounds with liberation of O_2	Pt, Pd, Co_3O_4
	A. III.	Complete oxidation of different substances:	
	A. III. 1.	Complete oxidation of different substances over metal catalysts	Pt, Pd, Ni
	A. III. 2.	Complete oxidation of different substances over oxide catalysts (and over salts)	Co_3O_4, $CuCo_2O_4$, $NiCo_2O_4$, MnO_2-CuO($CuCl$, $NiSn$, $CoSn$)
B. The processes greatly accelerated by catalysts involving compounds of elements belonging mostly to the Vb and VIb groups of the Mendeleev System (V, Mo)	B. I.	Oxidation of different organic compounds into acids or their anhydrides:	
	B. I. 1.	Oxidation of acyclic organic compounds into acids or their anhydrides	Complex catalysts on the basis of the highest oxides of molybdenum or vanadium (V_2O_5-MoO_3, Co_2O_3-MoO_3, etc.)
	B. I. 2.	Oxidation of side chains of the substituted aromatics or heterocyclic compounds into acids or their anhydrides (with the preservation of a cycle)	The highest oxides of vanadium (also with the additives of SnO_2, MoO_3 etc.)
	B. I. 3.	Oxidation of cyclic hydrocarbons and heterocycles into acids or their anhydrides (with the destruction of a cycle)	The highest oxides of vanadium (also with the additives of MoO_3, K_2SO_4, Fe_2O_3, Co_2O_3 etc.)
	B. II.	Ammonoxidation of cyclic hydrocarbons and heterocycles with the formation of nitriles	The highest oxides of vanadium (also with the additives of SnO_2, TiO_2, MoO_3 etc.)
	B. III.	Oxidation of acyclic alcohols into carbonyl compounds over oxide catalysts	Ferric - molybdenum oxide catalyst, the highest oxides of vanadium (also with the additives of Fe_2O_3, K_2SO_4 etc.)
	B. IV.	Oxidation of side chains of the substituted aromatics or heterocycles into carbonyl compounds (with the preservation of a cycle)	The highest oxides of molybdenum (also with the additives of oxides of vanadium, chromium, thorium, uranium etc.)
C. The processes greatly accelerated by catalysts involving compounds of elements belonging mostly to the VIb, IVa and Va groups of the Mendeleev System (Mo, W, Bi, Sn, Sb, P)	C. I.	Oxidative dehydrogenation of olefines on oxide catalysts	
	C. II.	Oxidative dehydrocyclization of olefines and their derivatives	Bismuth-molybdenum oxide catalyst
	C. III.	Ammonoxidation of olefines with the formation of unsaturated nitriles	
D. The processes greatly accelerated by catalysts involving elements or compounds of elements belonging mostly to the Ib group of the Mendeleev System (Cu, Ag)	D. I.	Oxidation of olefines (or diolefines) and their derivatives into unsaturated carbonyl compounds	Cuprous oxide a)
	D. II.	Oxidative dehydrogenation of alcohols with the formation of carbonyl compounds over metal catalysts	Silver, copper
	D. III.	Oxidative dehydrogenation of unsaturated amines with the formation of unsaturated nitriles	Silver, copper
	D. IV.	Oxidation of olefines into oxides of olefines b)	Silver

a) In the oxidation of olefines which cannot undergo oxidative dehydrogenation (propene, isobutylene), higher rates of
the formation of unsaturated carbonyl compounds are reached over the bismuth-molybdenum oxide catalyst than
over cuprous oxide, but only at considerably higher temperatures.

b) This type contains only one process (ethylene oxidation into ethylene oxide).

Nevertheless, these reactions include various initial substances and products, activity patterns and optimum catalysts which in all cases are almost the same. Inasmuch as this subclass also involves a reaction where oxygen is a single reagent (homomolecular oxygen exchange), it is natural to believe that within any catalytic series catalytic activity pattern is determined mainly by the affinity of catalysts to oxygen.

Fast rates of the above processes are likely to be assured by active surface oxygen which takes part in the formation of the limiting step transition state and possesses the optimum bond energy $(q_s)_{opt}$. It ensures fast oxygen adsorption as well as desorption in the reactions of oxygen exchange and liberation; in oxidation processes this active surface oxygen causes the easy rupture of chemical bonds in an oxidized molecule at all stages of the end-product formation.

This peculiarity determines the main requirement for optimum catalysts: they must have a relatively low oxygen bond energy, but this should not be too small to assure fast adsorption of oxygen.

The existence of rather clear correlations between the energy of the oxygen-catalyst bond (q_s) and the catalytic activity in complete oxidation, decomposition with O_2 liberation and oxygen exchange supports this conclusion.[4-9] The most active catalysts of the given reactions are Pt, Pd, Rh, Ni, Co_3O_4, MnO_2, NiO, CuO; the strength of oxygen bond for these substances is not high and at the same time their surfaces can rapidly attach oxygen under the conditions of catalysis.

The quantitative expression for the thermodynamic criterion of the optimum catalyst for the A-subclass reactions (i. e., the expression for $(q_s)_{opt}$) can be deduced by means of the Brønsted-Temkin relation

$$k = GK^\alpha$$

where k is the rate constant and K is the equilibrium constant for an elementary stage involving a catalyst; G and α are the values constant for catalytic systems of the same type.

However, the decisive role of the oxygen-catalyst bond energy in determining the relative catalytic activities does not mean that the properties of the other reagents (besides oxygen) are of no importance. Nevertheless, despite the similarity of the catalytic activity patterns for different processes of complete oxidation, the absolute rates of oxidation of various molecules over the same catalyst differ significantly. For example acetylenic hydrocarbons are usually oxidized better than olefines, and the latter react faster than paraffins over oxide catalysts;[9] carbon monoxide is oxidized faster than hydrogen in the presence of many oxides, and so on. This can be explained by a specific activation of oxidized molecules in the course of the formation of the limiting stage activated complex. It gives rise to apparent rates depending upon the energy of bonds breaking in these molecules and arising in their interaction with oxygen.

We presumed that in limiting steps of complete oxidation of hydrocarbons on oxide catalysts the breaking of carbon-carbon bonds in initial molecules (C-C-bonds in paraffins, π-C-C-bonds in olefines or acetylenic hydrocarbons) takes place. At the same time oxygen-containing radicals are formed on the surface and their further conversion into end-products proceeds readily. This hypothesis has been proved by the correlation between the energy of the above-mentioned bonds and the reactivity of hydrocarbons.[6]

On the other hand, the reactivity of the same molecules in complete oxidation over transition metals and oxide catalysts is often quite different. For instance the specific catalytic activity (W) of the optimum metal catalyst of hydrogen oxidation (Pt) is 4-5 orders higher than that of the optimum oxide catalyst (Co_3O_4).[10] Considerable differences are also observed in the rates of hydrocarbon combustion over metals and over oxides.[4] Hence the log $(W-q_s)$ correlation curves for metal and oxide catalysts are not matched, the Brønsted g and α constants for each being different.[4,6] This means the properties of the activated complexes are different. For that reason, complete oxidation processes over metal and oxide catalysts are attributed to different catalytic types (Table I).

These differences are most probably caused by unequal abilities of transition metals and oxides to activate oxidized molecules. This is sharply revealed in hydrogen oxidation. A high negative value of standard activation entropy of the rate-determining step of this reaction over the oxides of Co, Cr, Cu, Mn, Ni, V, Mo, W, Cd, Sn ($\Delta S_L^* = -38 \pm 2$ e. u.) shows the transition state to be localized.[11] An approximate constancy of the ΔS_L^* value means that the oxides are catalysts of the same type. On the other hand, the ΔS_L^* value for hydrogen oxidation over transition metals (Pt, Pd, Ni) is considerably less negative (-16 ± 2 e. u.), which suggests that in the latter case an activated complex is mobile.[11] The hydrogen adsorbed with dissociation over Pt and Pd is known to migrate along the surface,[12,13] and this mobility is probably kept in transition state.

Thus one can see that in interpreting the classification, the correlations between catalytic and thermodynamic properties of substances as well as the analysis of entropy activation values are very useful.

One should discriminate between the ΔS_L^* values of activation entropy ΔS^*_{eff} calculated from experimental rate constants and apparent activation heats E_{eff}. While the former is nearly constant for catalytic systems of the same type, the latter can vary greatly in some cases. These variations are usually succeeded by parallel changes in the E_{eff} values so that the "Compensation Effect" (CE) is observed:

$$\log (K_0)_{eff} = \gamma E_{eff} + C$$

($(K_0)_{eff}$ is the effective preexponential factor which is proportional to $\exp(\Delta S^*_{eff} /R)$; γ and C are constants).

We[14] have demonstrated that in most cases the ΔS^*_{eff} variations and the apparent CE are caused by variations in the reaction kinetic orders.

For example, Fig. 1, a shows the experimental $\log (K_0)_{eff}$ - E_{eff} dependence for complete oxidation of propene over oxides.[9] This reaction obeys the kinetic equation

$$W = P_{C_3H_6}^m \cdot P_{O_2}^n$$

where the m and n values are variable. According to ref. 15, $\Delta S^*_{eff} \approx \Delta S_L^* + m\Delta S^0_{C_3H_6} + n\Delta S^0_{O_2}$, where $\Delta S^0_{C_3H_6}$ and $S^0_{O_2}$ are standard adsorption entropies. Assuming ΔS_L^* to be constant, one should expect the $m \log (\sigma_0)_{C_3H_6} + n \log (\sigma_0) O_2$ — E_{eff} dependence to be similar to that shown in Fig. 1a $((\sigma_0)_{C_3H_6}$ and $(\sigma_0)_{O_2}$ are the preexponents in the expressions for adsorption coefficients, the former being equal to $\exp (\Delta S_i^*/R)$; our calculations that $(\sigma_0)_{C_3H_6} = (\sigma_0)_{O_2} = 10^{-9}$ atm^{-1}, which corresponds to the localized adsorption of the reagents). The nearly equal slope of the straight lines

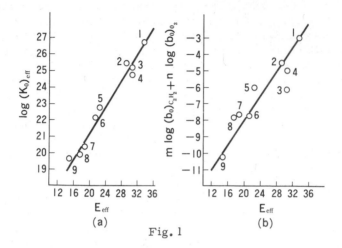

Fig. 1

(a)   (b)

shown in Fig. 1, a and b (the equality of the γ-coefficients)
means the CE to be related mainly to the m and n variations. This result also proves the approximate constancy of
the ΔS_L^* values.

Thus if one eliminates the distorting influence of such an
apparent CE for catalytic systems of the same type, the
activation entropy of the limiting step turns out to have an
approximately constant value. Similar results were obtained
for some other reactions of complete oxidation.[4]

The processes of partial oxidation, which form three
extensive subclasses (B, C, D), are more complex. The
diversity of corresponding catalytic types and optimum
catalysts show the existence of several types of partial
oxidation mechanisms.

The requirements for optimum catalysts of partial oxidation are more complicated than that of complete oxidation
because as well as having the appropriate activity, it must
first of all be highly selective.

In view of this, such a catalyst should form those adsorption complexes which are readily converted into the desired
products. However, it is insufficient merely to ensure high
selectivity. It is also necessary that a given catalyst should
accelerate only poorly any possible transformations of
original molecules and partial oxidation products in undesirable directions. The most undesirable conversion is complete oxidation which is usually preferable thermodynamic-
ally.

The latter limitation allows us to conceive of certain regularities observed in partial oxidation. As the catalysts of these reactions should not actively accelerate complete combustion, it is necessary to use substances not possessing high total activities, which must therefore act at high temperatures. Most partial oxidation processes occur over the best catalysts at higher temperatures compared with complete oxidation of the same molecules over the corresponding optimum catalysts. Examples of this are given in Table II.[3,4]

As an example showing general peculiarities of partial oxidation, we shall consider the processes of oxidation of various organic compounds into acids or their anhydrides.[16]

According to our classification, these reactions belong to a single subclass because their acceleration used the substances of very similar chemical nature: either the highest oxides of vanadium (individual or with additives) or the highest oxides of molybdenum with additives. Typical reactions of this subclass are given in Table III. Among them are processes of commercial significance, such as benzene oxidation into maleic anhydride and naphthalene or ortho-xylene oxidation into phthalic anhydride.

In order to obtain a partial oxidation product a fairly high reacting rate should be ensured. The reaction product must be stabilized on a surface so as to prevent its further oxidation.

Evidently the realization of these requirements depends to a considerable extent on the oxygen-catalyst bond energy. According to the q_s values (heats of dissociation without the lower oxide phase formation, kcal[6]) simple oxides can be divided into the following three groups:

(a) Ag_2O, Co_3O_4, MnO_2, NiO, CuO ($q_s < 35$)

(b) Fe_2O_3, Fe_3O_4, FeO, CdO, SnO_2, V_2O_5
     (V_6O_{13}, Cr_2O_3)   ($q_s = 50\text{-}70$)

(c) UO_3, ZnO, CeO_2, WO_3, MoO_3, TiO_2 (Sb_2O_5) ($q_s > 80$)

Similar combinations can be obtained on the basis of initial heats of oxygen desorption[5] or activation heats (E_0) of oxygen isotopic exchange.[5,17]* The (a)-group oxides (Co_3O_4, MnO_2, NiO, CuO) are characterized by relatively low q_s values which are close to the optimum q_s values for

* On the basis of the E_0 values chromium oxide has been placed in the (b)-group and Sb_2O_5- in the (c)-group (the q_s data for these oxides are absent).

Table II

Complete and Partial Oxidation of the Same Molecules over the Corresponding Optimum Catalysts.

Oxidized Molecules	Complete Oxidation	Partial Oxidation
Ethylene	Pt, 5-100°c	Into ethylene oxide: Ag, 180-220°c
Propene	Pt, 80-150°c Co_3O_4, 215-240°c	Into acrolein: Cu_2O, 300-380°c
Pentene-2	Co_3O_4, 180°c	Into pentadiene: $Bi_2O_3 \cdot 2MoO_3$, 460°c
Benzene	Co_3O_4, 170°c	Into maleic anhydride: V_2O_5-MoO_3, 400-500°c
Benzaldehyde	MnO_2, 180°c	Into benzoic acid: V_2O_5-SnO_2, 310-340°c

Table III

Examples of Heterogeneous Catalytic Reactions of Oxidation
of Organic Substances into Acids or their Anhydrides.

Oxidized Substances	Products	Catalysts
Acrolein	Acrylic acid	Co_2O_3-MoO_3, SnO_2-MoO_3, V_2O_5-MoO_3-Al_2O_3
Butene-1	Maleic anhydride	V_2O_5-P_2O_5, Co_2O_3-MoO_3, Co_2O_3-MoO_3-P_2O_5, Fe_2O_3-MoO_3
Benzene	Maleic anhydride	V_2O_5, V_2O_5-MoO_3, V_2O_5-SnO_2
o-Xylene	Phthalic anhydride	$V_2O_5(V_6O_{13})$
Phthalic anhydride	Maleic anhydride	V_2O_5
Benzaldehyde	Benzoic acid	V_2O_5, V_2O_5-SnO_2
Durene	Dianhydride of pyromellitic acid	V_2O_5, V_2O_5-P_2O_5
Naphthalene	Phthalic anhydride	V_2O_5, V_2O_5-K_2SO_4-SiO_2, V_2O_5-SnO_2, V_2O_5-TiO_2
Acenaphthene	Naphthalic anhydride	V_2O_5, V_2O_5-K_2SO_4/pumice, Fe_2O_3-V_2O_5-K_2SO_4/pumice
Furfural	Maleic anhydride	V_2O_5, V_2O_5-MoO_3

complete oxidation of different organic substances; for this reason these oxides shouldn't be suitable as partial oxidation catalysts.

Indeed, the mentioned oxides actively catalyze the full combustion of olefines, benzene, toluene, naphthalene, benzaldehyde and other organic molecules. Silver oxide has a very small affinity to oxygen so that oxygen activation over this catalyst should be difficult. The (c)-group oxides have such high qs values that great difficulties should be expected interacting an oxidized molecule with adsorbed oxygen over these oxides, resulting in extremely slow rates of initial molecule oxidation.

The most suitable oxides should be those of the (b)-group, for which the q_s values are between those of the (a)- and (c)-groups.

However, for partial oxidation rather than complete oxidation the favorable oxygen-catalyst bond energy is not sufficient to produce optimum conditions.

We have made the following general assumption: the structure of the activated complexes of the slowest steps of partial oxidation is close to that of the reaction products. Because partial oxidation products in the case under consideration are acids or their anhydrides, it is natural to think that the transition state structure in these reactions is similar to salt-like surface complexes. Consequently, corresponding active and selective catalysts should be able to form such complexes which are readily transformed into end-products.

This ability is determined by the acid-base properties of catalytic surfaces, the most favorable being amphoteric properties which provide rather high formation and decomposion of the aforesaid complexes. Under reaction conditions the catalysts are probably lower than under oxidation states. Therefore, one should take the acid-base characteristics into account not only for an initial catalyst, but its reduced form.

If we apply simultaneously both criteria of optimum conditions (oxygen-catalyst bond energy and acid base properties of a catalyst), we must rule out the oxides of cadmium, tin and iron from the (b)-group: the Cd^{2+}, Sn^{2+}, Fe^{2+}, Cr^{2+} cations give oxides with mainly base properties. These oxides can form rather stable salts of organic acids unfavorable for catalysis. Thus only the higher acidic oxides of vanadium have remained.

Their reduced form corresponds to the amphoteric oxide
V_2O_4; hence in interactions of organic molecules with vanadium oxide catalysts, unstable salt-like compounds should
appear. In the case of naphthalene[18] and benzaldehyde[7]
such compounds have really been detected by IR-spectros-
copy.

Now we can see that among simple oxides only the highest
oxides of vanadium satisfy simultaneously both criteria for
optimum conditions. We have shown that those complex
catalysts with a basis of V_2O_5 or MoO_3 also satisfy
qualitatively the optimum criteria of acid-base and redox
properties[4,16] This explains the use of these substances
as the best catalysts of partial oxidation of different organic
compounds into acids or their anhydrides.

It is necessary to note the difference between partial
oxidation of acyclic compounds and aromatics which is
reflected by the classification. While for acyclic compounds
either vanadium oxide or molybdenum oxide catalysts are
used, for aromatic only vanadium oxide catalysts are suitable. This may be explicable by the strong necessity for
multipoint adsorption of aromatics in the course of partial
oxidation. This would result in the optimum catalysts of
these reactions having to satisfy not only the energetic
requirements but also the requirements of structural
(geometric) correspondence between an oxidized molecule
and the lattice of a catalyst. According to ref. 19, the O-O
distances in the elementary cells of V_2O_5 (and V_6O_{13}) are
very close to the C-C distances in benzene and in other
aromatics.

Let us suppose that in the oxidation of aliphatic alcohols
into aldehydes over oxides the dissociative adsorption of
alcohols takes place leading to the formation of salt-like
surface structures of alcoholate type[20] Then the catalysts
of these processes should have the qualitative requirements
for optimum condition which we have deduced for acid
formation catalysts. Hence it becomes comprehensible why
among individual oxides the greatest activity in partial
oxidation of alcohols is shown by the highest oxides of vanadium and why complex catalysts used in practice are the
salts of molybdenum acid.

In this way we can explain why partial oxidation of organic
substances into acids and partial oxidation of alcohols into
carbonyl compounds belong within the same catalytic subclass. It is due to the similar nature of the intermediate

chemical interaction of reagents with catalysts which leads
to the resemblance of requirements for optimum catalysts.

At first sight there is a close similarity between the
partial oxidation of organic substances into acids and the
oxidation of SO_2 to SO_3. Indeed, the product of the latter
reaction is the anhydride of a strong acid and the best oxide
catalyst of this reaction is also vanadium pentoxide.
However, this process belongs to complete oxidation for
which it is not selectivity but rather high activity which is
important. The highest rates of complete oxidation are
reached over the catalysts characterized by relatively low
oxygen-catalyst bond energy while optimum catalysts of
partial oxidation have elevated q_s values. Accordingly,
vanadium pentoxide (for which the q_s value is relatively
high)[4-6] is a true optimum (or almost optimum) catalyst for
partial oxidation of organic substances into acids, but this
oxide must not be regarded as a true optimum catalyst of the
oxidation of SO_2 to SO_3. The V_2O_5 used in this process
because oxides with q_s values close to $(q_s)_{opt}$ (Co_3O_4, MnO_2,
NiO, CuO) turn into inactive sulphates under the conditions
for sulfuric acid catalysis. Thus in the SO_2 oxidation the
strong acidic properties of a product only confine the choice
of real catalysts, though the optimum catalyst criterion does
not differ from that of other processes of complete oxidation.
This conclusion is supported by the fact that the SO_2 oxidation is perfectly accelerated by metallic platinum for which
the q_s value is very close to the $(q_s)_{opt}$. value for complete
oxidation. At the same time platinum is usually not used in
gas phase catalysis of partial oxidation of organic substances.

The following conclusion can be made on the basis of our
analysis. Optimum catalysts of the considered partial
oxidation processes should satisfy simultaneously several
criteria of optimum which can not be reduced to a single
criterion of the optimum oxygen-catalyst bond energy.
Therefore, generally speaking, we must not expect clear
correlations between the q_s values and catalytic activity with
respect to these processes, in contrast to complete oxidation. Correlations between the q_s values and catalytic
properties of substances with respect to partial oxidation
should be searched only within the few groups of catalysts
for which other characteristics of stability of the assumed
intermediates (for instance, acid base properties in the
case of acids formation) are close.

An example of such a group of catalysts is offered by the

vanadium-molybdenum oxide catalysts investigated in our
work.[21]   As the measure of surface oxygen bond strength we
used equilibrium pressure of desorption of O_2 (Po).   This
was measured by a procedure similar to ref. 5.   Figure 2
shows a satisfactory correlation between oxygen bond
strength and catalytic activity with respect to benzene oxidation into maleic anhydride.   A similar correlation is
observed for methanol oxidation into formaldehyde because
the activity patterns in this reaction[22] and in partial benzene
oxidation[23] are nearly the same.

   It is interesting to show a definite correlation between the
character of the processes of a given subclass and the position in the Mendeleev System of the elements which form
the corresponding optimum catalysts.

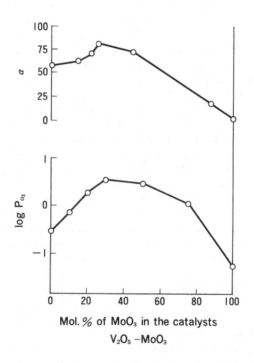

Mol. % of MoO$_3$ in the catalysts

V_2O_5 −MoO$_3$

Fig. 2

The first subclass (A) processes involving oxygen isotopic exchange, decomposition with O_2 elimination and complete oxidation are actively accelerated mainly by the
catalysts on the basis of elements or the compounds of
elements of group VIII (Pt, Pd, Co, Ni). The second
subclass (B) reactions, which may be regarded as rather
"heavy" partial oxidation reactions (the formation of organic
acids, their anhydrides, nitriles, often demanding the
carbon-carbon bond rupture) are usually catalyzed by the
substances containing the highest oxides of elements of
groups Vb or VIb (V, Mo). The best catalysts of the third
subclass (C) partial oxidation (oxidative dehydrogenation
and dehydrocyclization, ammonoxydation) contain the higher
oxides of the VIb group elements (Mo, W) together with the
oxides of group IVa and Va elements (Bi, Sb, P, Sn). The
fourth (D) subclass includes "mild" partial oxidation
reactions, mostly accelerated by the catalysts containing
group Ib elements (Cu, Ag).

These subclasses cover the main body of experimental
data concerning catalytic properties of solids in gas phase
reactions involving O_2. However, some processes have not
been included in our classification. The peculiarities of
some of them do not allow us to group them with heterogeneous catalytic reactions; the rest have been poorly studied.

The first type processes include heterogeneous homogeneous reactions[24] such as the partial oxidation of paraffins
or methylacetylene and the oxidation of methane-ammonia
mixtures to form HCN, which proceeds at extremely high
temperatures ($\sim 1,000°C$) etc.

Also, the recombination of O-atoms has been ruled out
because its rate depends not only on intrinsic catalytic
properties of a solid but also on the ability of the latter to
comsume the liberating energy in order to prevent O_2
decomposition. Therefore, this reaction is catalyzed not
only by the transition metal oxides but also by the halides of
alkalis and alkaline earth metals, which are known to be
inactive in oxidation catalysis.

The second type reactions (inadequately examined) are
such processes as the oxidative dehydrogenation of hydrocarbons over metals, destructive oxidation with the formation of carbonyl compounds, oxidative condensation,
oxidative dealkylation etc.

The proposed classification has not been completed and

should be further developed as new experimental data appear. This classification is relative to some extent. The limits of catalytic types and especially that of subclasses are not rigid. The division of solids into partial and complete oxidation catalysts is rather relative. At elevated temperatures over many partial oxidation catalysts organic molecules undergo full combustion.

Strictly speaking, the concept of the "catalytic system" must involve the reaction conditions. A change in these can result in the alteration of the mechanism followed by changes in catalytic activity patterns and in other regularities important for classification.

At elevated temperatures and near explosion limits heterogeneous catalytic processes are likely to proceed as heterogeneous homogeneous ones. So, in studying hydrogen oxidation over V-Mo-oxide catalysts under small O_2 excesses in mixtures (near explosion limits on concentration)[25] we have found all peculiarities of heterogeneous homogeneous catalysis. Under these conditions there is no correlation between catalytic activity and the surface oxygen bond energy. With great excesses of O_2, homogeneous steps drop sharply, and the above correlation is observed.

Homogeneous stages must be taken into account in further development of the classification. Typical heterogeneous homogeneous reactions are to be placed in special class lying between heterogeneous catalytic oxidation and homogeneous chain reactions of oxidation.

Some heterogeneous catalytic reactions involving O_2 are likely to belong to more general catalytic classes. For solving this problem the development of a complete classification of catalytic reactions is necessary.

Nevertheless bearing in mind these limitations, the proposed classification can be used now in creating the theory of prediction of catalytic action. The elucidation of the rational basis of the classification leads to certain conclusions about the peculiarities of mechanisms of various processes and allows us to formulate qualitative requirements for the corresponding optimum catalysts.

At present this classification may be useful in practice because it can serve as a guide in improving catalysts. So, if some reactions are combined in a single catalytic type or subclass, the experience of selection and improvement of catalysts in well studied reactions can be applied to insufficiently studied ones.

The direct way of verifying the classification is by detailed research into the mechanisms of catalytic reactions. From this point of view, the given classification can help to choose the most interesting and important objects for experimental study.

References

1) V.A. Roiter, Ukrainski Khim. Zsurnal, 19, 119 (1953); Kataliz i katalizatori, 1, 5 (1965); Kinetika i kataliz, 8, 1034 (1967); Z.V.A. Roiter and G.I. Golodets, Problemi kinetiki i kataliza, 11, 56 (1966).

2) Catalytic Properties of Substances. Handbook (V.A. Roiter, ed.), Naukova Dumka, Kiev, vol. 1 (1968), vol. 2 (in press).

3) G.I. Golodets and Yu.I. Pyatnitzky, Kataliz i Katalizatori, 6, 5 (1970); Dokl. AN USSR, No.5, 416 (1969).

4) G.I. Golodets, Dokt. Dissertation, Inst. Phys. Chem. Akad. Sci. Ukrain. SSR, Kiev (1969).

5) G.K. Boreskov, V.V. Popovsky, and V.A. Sazonov, in Proceedings 4th Int. Cong. Catalysis, Nanka, Moscow, 1970, vol. 1, p. 343.

6) V.A. Roiter, G.I. Golodets, and Yu.I. Pyatnitzky, ibid., p. 365.

7) W.M.H. Sachtler, G.J.H. Dorgelo, J. Fahrenfort, and R.J.H. Voorheve, ibid., p. 355.

8) D.G. Klissurski, ibid., p. 374.

9) Y. Moro-oka and A. Ozaki, J. Catalysis, 5, 116 (1966); Y. Moro-oka, Y. Morikawa and A. Ozaki, J. Catalysis, 7, 23 (1967).

10) V.V. Popovsky and G.K. Boreskov, Problemi Kinetiki i Kataliza, 10, 67 (1960); E.N. Kharkovskaya, G.K. Boreskov, and M.G. Slinko, Dokladi Akademii Nayk, 127, 145 (1955).

11) V.V. Goncharuk, G.I. Colodets, and V.A. Roiter, Teoretitcheskaya i Eksperimentalnaya Khimia, 4, 688 (1968); Kataliz i Katalizatori, 7, 19 (1971).

12) N.I. Il'chenko and V.A. Yuza, Kataliz i katalizatori, 2, 118 (1966);

13) N.I. Il'chenko, Uspekhi Khimii, 40, 12 (1971).

14) G. I. Golodets, V. V. Goncharuk, and V. A. Roiter,
     Teoretitcheskaya i Eksperimentalnaya Khimia, 5,
     201 (1969).
15) G. I. Golodets and V. A. Roiter, Kinetika i Kataliz,
     4, 173 (1963).
16) G. I. Golodets, Dokladi Akademii Nayk CCCP, 172,
     133 (1967).
17) A. I. Gelbstein, S. S. Stroeva, Yu. M. Bakshi, and
     Yu. A. Mishchenko, in Proceedings 4th Int. Cong.
     Catalysis, Nauko, Moscow, Vol. 1, p.251.
18) B. M. Odrin and L. M. Roev, Kataliz i
     Katalizatori, 2, 133 (1966).
19) T. Vrbaški and W. K. Mathews, J. Catalysis, 5,
     125 (1966).
20) P. Jirú, M. Krjivanek, I. Novakova, and
     B. Vicherlova, Mechanism and Kinetics of Complex
     Catalytic Reactions, (Symposium at the 4th Int.
     Cong. on Catalysis), Moscow, 1968, paper No. 19.
21) Yu. I. Pyatnitzky, N. I. Il'chenko, and G. I. Golodets,
     Kataliz i Katalizatori, 7, 5 (1971).
22) L. N. Kurina and L. G. Maydanovskaya, Problemi
     Kinetiki i Kataliza, 16, 216 (1970); G. I. Golodets,
     Yu. I. Pyatnitzky, and N. I. Il'chenko, Dokladi
     Akademii Nayk CCCP, 196 579 (1971).
23) I. I. Ioffe, Z. I. Ezhkova, and A. G. Lubarsky,
     Kinetika i Kataliz, 14, 216 (1970).
24) M. V. Polyakov, Uspekhi Khimii, 17, 351 (1948).
25) N. I. Il'chenko, G. I. Golodets, and Yu. I. Pyatnitzky,
     Kinetika i Kataliz (in press).

Studies of Heterogeneous Catalysis by the Pulse Reaction Technique

Y. Murakami

Nagoya University, Nagoya

INTRODUCTION

An immediately apparent advantage of the pulse reaction technique is that a great deal of data can be obtained in a very short time compared with the time required using a steady-state flow reactor system. Another advantage is that the technique allows one to study more exhaustively initial rates, poisoning effects, catalyst deactivation, and other relaxation phenomena which often give insight into the reaction mechanism of the catalyst. The pulse reactor system, therefore, is now in use in many laboratories, but primarily as a qualitative analytical tool.

An important difference between the data resulting from continuous flow and pulse reactions is attributable to the fact that in the former after steady-state is reached a reactant exists at a constant partial pressure in each part of catalyst bed, while in the latter the pulse of reactant gets through the bed steady-state is attained.

In this paper the potential applications of the pulse reaction technique to studies on the mechanism of heterogeneous catalytic reactions are discussed on the basis of examples.

THEORY

The main feature of schemes for catalytic reactions is that an intermediate complex is formed between the catalyst and at least one of the substrate molecules, and that this complex undergoes subsequent reactions:

$$A + S \underset{k_{-1}}{\overset{k_1}{\rightleftharpoons}} AS \tag{1}$$

$$AS \xrightarrow{k_2} p + S \tag{2}$$

203

where A is a reactant molecule, P a product molecule, S a surface site, AS a complex adsorbed on the surface, and k_1, k_{-1}, and k_2 the rate constants for adsorption, desorption, and reaction, respectively.

Adsorption in Equilibrium

Equilibrium adsorption is essential if the rate of surface reaction is very slow compared with the rates of adsorption and desorption of reactant, and if a negligible fraction of the surface is covered by adsorbed product. We illustrate this case for a unimolecular reaction in which the Langmuir adsorption equilibrium for the reactant is

$$C_{AS} = \frac{KC_A}{1 + KC_A} \qquad (3)$$

where $k = k_1/k_{-1}$ is the equilibrium constant of adsorption, and C_A and C_{AS} represent concentrations (mole/cm^3 of catalyst bed).
The rate of reaction is then

$$r = k_2 C_{AS} = \frac{k_2 KC_A}{1 + KC_A} \qquad (4)$$

(1)   Weak adsorption

If adsorption is very weak and KC_A can be neglected compared with unity, Eq. (4) becomes Eq. (5):

$$r = k_2 KC_A \qquad (5)$$

Since this is an equation for a linear reaction, the conversion of reactant and the yield of products are in agreement with those of the flow system.

(2)   Strong adsorption

If unity can be neglected compared to KC_A in the denominator of Eq. (4), because adsorption is strong, Eq. (4) becomes Eq. (6):

$$r = k_2 \qquad (6)$$

In this case the problems are the same as in Section "Irreversible Adsorption" (2).

(3)   Moderate adsorption

When adsorption strength is moderate, the relations of conversion and $\varDelta X$ to the length of catalyst bed are shown in Figs. 1 and 2, respectively. $\varDelta X$, which represents the

deviation from the flow reaction, is defined by Eq. (7) and
is equivalent to zero in the flow reaction and to unity in the
case of the linear isotherm:

$$\Delta X = \frac{X - X_F}{X_L - X_F}$$ (7)

where X_F is the conversion with the Langmuir isotherm and
X_L is the conversion with the linear isotherm by the flow
reaction. As shown in Fig. 1, the conversion in the pulse
reaction lies between that in the flow reaction (lower dotted
line) and that in the linear isotherm (upper dotted line), and
the pulse width has a great effect on the conversion. The
concentration profiles at the end of bed are presented in
Figs. 3 and 4, where the shorter the pulse width is, the
lower the concentration in the bed becomes. In the reaction
A→R, the difference of the pulse and flow reactions is
mainly due to the lowering of the concentration, but the
separation between components has little effect. Even when

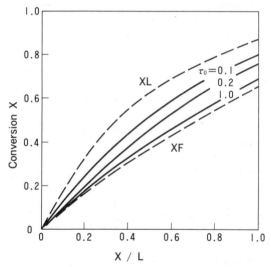

Fig. 1. Conversion as a function of bed length at different
pulse widths in A→R (adsorption in equilibrium).

$k_1 k_A \theta = 2$, $K_A = 14$, $K_R = 0.5$, $\lambda = 0.1$.

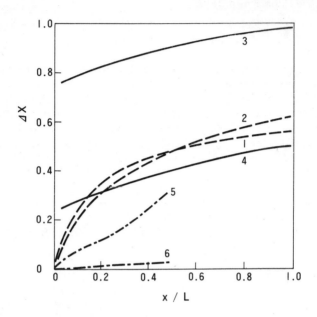

Fig. 2. Deviation from the flow technique. $\lambda = 0.1$.
Curves 1 and 2:
A→R (adsorption in equilibrium)

$k_1 k_A \theta = 2$, $\tau_0 = 0.1$, $k_A = 5$, (1) $K_R = 0.5$
(2) $K_R = 5$.

Curves 3 and 4:
A→R (adsorption in nonequilibrium)

$k_1 k_A O = 2$, $k_{aA} O = 4$, $K_A = 5$, $K_R = 0.5$,
(3) $\tau_0 = 0.1$, (4) $\tau_0 = 0.5$.

Curves 5 and 6:
A⇄R (adsorption in equilibrium)

$k_1 K_A O = 2$, $K - 0.5$, $K_A = 50$, $K_R = 5$,
(5) $\tau_0 = 0.1$, (6) $\tau_0 = 1$.

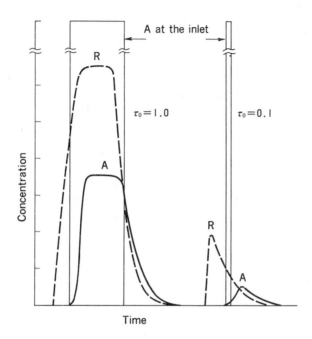

Fig. 3.   Concentration profiles at the end of the catalyst
bed in A→R (adsorption in equilibrium).

$$k_1 K_A \theta = 2, \quad K_A = 14, \quad K_R = 0.5, \quad \lambda = 0.1.$$

$K_A = K_R'$ that is, the adsorption strength of A is equivalent to
that of R, the conversion in the pulse reaction is greater than
that in the flow reaction (curve 2 in Fig. 2). It is important
to note that if the adsorption isotherm is linear, that is, in
the case of Section "Weak adsorption," there is no difference
between the pulse and the flow reactions in the reaction A→R.
Then the conversion increases with shortening the pulse
width τ, until at $\tau_0 = 0$ it becomes equal to that in the linear
adsorption isotherm (Fig. 1).

(4)   Bimolecular reaction

When two kinds of reactants, A and B, react, both
reactants are adsorbed with different strengths and are
transferred at different speeds along the catalyst bed.

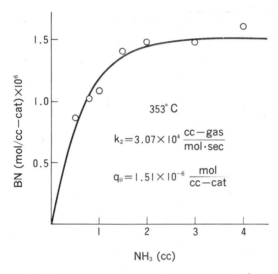

Fig. 4. Concentration profiles at $x = L/2$ in A-R.

$$k_1 K_A \theta = 2, \quad K_A = 5, \quad K_R = 0.5, \quad \lambda = 0.1,$$
$$\tau_0 = 0.1.$$

The separation of reactant zones is of great disadvantage to the reaction. This tendency becomes more pronounced as pulse width becomes shortened. In the case of a reaction between two molecules of A, the conversion of A is constant, regardless of pulse width, since there is no separation of zones.

Adsorption in Nonequilibrium

If surface reaction rates cannot be neglected in comparison with rates of adsorption and desorption, then a steady-state nonequilibrium treatment may be employed in the flow reaction alone. We have

$$\frac{dC_{AS}}{dt} = k_1 C_A C_S - k_{-1} C_{AS} - k_2 C_{AS} = 0 \qquad (8)$$

The reaction rate is

$$r = k_2 C_{AS} = \frac{k_2 K' C_A}{1 + K' C_A} \tag{9}$$

$$K' = k_1/(k_{-1} + k_2)$$

This is of the same form as Eq. (4), if K' is taken to be $k_1/(k_{-1} + k_2)$ rather than as k_1/k_{-1}. Therefore, neither of the reaction schemes can be distinguished by the flow reaction technique.

In the pulse reaction, ΔX from the nonequilibrium adsorption mechanism is larger than that from the equilibrium adsorption mechenism, as shown in Fig. 2.

Irreversible Adsorption

If k_{-1} is negligible in Eq. (9), removal of the adsorbed complex AS is by reaction (2) alone. In the pulse reaction the peak of A does not spread because there is no desorption, and it keeps its initial pulse width.

(1) $k_1 \ll k_2$

The adsorbed reactant AS is consumed immediately by reaction (2) and the concentration of AS on the surface can be neglected. The overall reaction rate is determined by step (1). In the flow and pulse reactions, the reaction rate is

$$r = k_1 C_A \tag{10}$$

(2) $k_1 \gg k_2$

In the flow reaction by the steady-state treatment the reaction rate is

$$r = k_2 \tag{11}$$

The pulse reaction situation is much more complex. While pulse A exists in the gas phase in the bed, AS is formed at the rate $k_1 C_A (C_0 - C_{AS})$, after which it decomposes at the rate $k_2 C_{AS}$. (C_0 is the total concentration of the active site).

If a sufficient amount of A is injected, almost all the active sites may be covered by AS, so we can know the number of active sites from the amount of decomposition product of AS. Furthermore, the rate of surface reaction, i.e., of decomposition of AS, can be measured by analysis of tailing, and represents the rate of surface reaction at corresponding time. After passage of pulse A, the rate equation is

$$\frac{dC_p}{dt} = -\frac{dC_{AS}}{dt} = k_2 C_{AS} \tag{12}$$

Integrating this equation, using the fact that $C_{AS} = C_0$ at t=0, gives

$$\ln C_{AS} = \ln C_0 - k_2 t \tag{13}$$

Expanding Eq. (12) as the logarithm, and substituting Eq. (3), we get Eq. (14):

$$\ln\left(\frac{dC_p}{dt}\right) = \ln k_2 C_0 - k_2 t \tag{14}$$

where (dC_p/dt) corresponds to the height of the chromatographic peak tailing of product.

Polystep Reactions
    As an example we take up a reaction scheme:

$$A + S \xrightarrow{K_1} AS \tag{15}$$

$$AS + B \xrightarrow{k_2} R + S \tag{16}$$

Under steady-state conditions the rate equation for the flow reaction is

$$\frac{C_0}{r} = \frac{1}{k_1 C_A} + \frac{1}{k_2 C_B} \tag{17}$$

Assuming first order kinetics for the both of concentrations of gas component and surface site, we get Eqs. (1) and (2) for step (15):

$$\frac{\partial C_{AS}}{\partial t} = k_1 C_A \left(1 - \frac{C_{AS}}{C}\right) \tag{18}$$

$$\frac{\partial C_A}{\partial t} + u\frac{\partial C_A}{\partial X} + \frac{\partial C_{AS}}{\partial t} = 0 \tag{19}$$

Initial condition: $C_A = 0$ and $C_{AS} = 0$ at $t = 0$;
boundary condition: $C_A = C_{AO}$ at $x = 0$;

$$C_{AS} = 0 \text{ at } x = ut$$

From Eq. (18) we get Eq. (20) at x = 0:

$$C_{AS} = C_0 \{1 - \exp(-k_1 C_{A0}t/C_0) \} \tag{20}$$

From Eqs. (18) and (19) we get Eq. (21) at x=ut:

$$C_A = C_0 \exp(-k_1 \theta_X) \tag{21}$$

where $\theta_X = x/u$.

The solutions of Eqs. (18) and (19) are Eqs. (22) and (23):

$$C_{AS} = C_0 \frac{[1 - \exp\{-(k_1 C_{A0}/C_0)(t - \theta_X)\}] \exp(-k_1\theta_X)}{1 - [1 - \exp\{-(k_1 C_{A0}/C_0)(t - \theta_X)\}][1 - \exp(-k_1\theta_X)]} \tag{22}$$

and

$$C_A = C_{A0}\frac{\exp(-k_1 \theta_X)}{1 - [1 - \exp\{-(k_1 C_{A0}/C_0)(t-\theta_X)\}][1 - \exp(-k_1\theta_X)]} \tag{23}$$

If a rectangular pulse with a concentration C_0 and width t_0 is injected, Eq. (22) is held at $t - \theta_X \leq t_0$ and $C_A = 0$ and $C_{AS} = C_{AS}(t_0 + \theta_X)$ at $t - \theta_X > t_0$. The surface concentration profile after passage of the pulse can be obtained by substituting t_0 for $t_0 - \theta_X$:

$$C_{AS} = C_0 \frac{\alpha}{\alpha + \exp(k_1 \theta_X)} \tag{24}$$

$$\alpha = \exp(k_1 C_{A0}t_0/ C_0) - 1 \tag{25}$$

$$= \exp(k_1 \theta \, \Omega / C_0 V) - 1 \tag{26}$$

where $\Omega = C_{A0}t_0 V/\theta$ is a pulse size injected, $\theta = V/F$, F = flow rate of carrier gas, and V = volume of catalyst bed. The mean concentration of AS (mol/ml-bed) in the bed is given by Eq. (27):

$$\overline{C}_{AS} = \frac{1}{L} \int_0^L C_{AS}dx$$

$$= C_0 \left[1 - \frac{1}{k_1 \theta} \ln \frac{\alpha + \exp(k_1\theta)}{\alpha + 1} \right] \tag{27}$$

We can estimate the rate constant k_1 and the total concentration of the active site C_0 by plotting $\overline{C}_{AS}$ against Ω at various catalyst volumes and using Eqs. (26) and (27). These equations can be used for any shape of inlet pulse. The rate constant of step (16), k_2, can be estimated in a similar way.

The pulse reaction technique has the advantage that the rate constant per active site of each individual step of a reaction containing more than one reactant can be obtained.

EXPERIMENTAL RESULTS

Disproportionation of Cumene

The yield of diisopropylbenzene is independent of pulse width, but depends on the initial partial pressure of cumene, as shown in Fig. 5. It follows that diisopropylbenzene (DIB) is considered to be formed by the order of DIB formation is larger than unity for cumene. If it is formed in the alkylation of cumene with propylene (C+P=DIB), the yield of DIB will

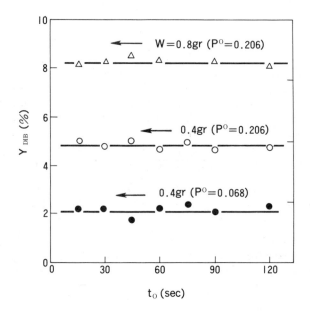

Fig. 5. Dependence of yield of diisopropylbenzene on pulse width. Reaction temperature is 300°C.

decrease with decreasing pulse width because of the separation between cumene and propylene. When the pulse width is
sufficiently small, most of propylene is separated from
cumene, and the alkylation does not proceed because propylene
once separated from cumene cannot react any more.
However, such dependence of the yield of DIB on the pulse
width was not observed experimentally. As mentioned above,
the yield of DIB is independent of the pulse width and furthermore, even when the pulse width is about a few seconds (such
small pulse width can be obtained by injecting cumene with a
microsyringe), a considerable amount of DIB is obtained.
For that reason it was concluded that DIB is not a product of
the alkylation of cumene with propylene.

Dehydration of Ethanol over an Alumina Catalyst
    This is a typical example of irreversible adsorption of a
reactant followed by slow surface reaction (Section "Irreversible Adsorption" (2)). A chromatogram obtained by
the pulse reaction of ethanol is shown in Fig. 6. The ethylene
peak has abnormally long tailing, but the ethanol and ether
peaks are sharp. A plot of the logarithmic heights of the
ethylene peak (which correspond to $\ln(dC_p/dt)$ in Eq. (14))
against time on Eq. (14) was found to be linear. The
surface reaction rate was obtained from the slopes of

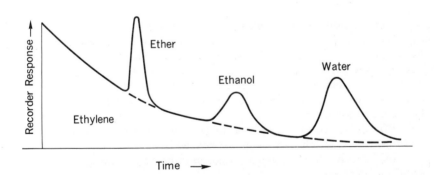

Fig. 6. Typical chromatogram of the pulse reaction of
ethanol over an alumina catalyst. Reaction temperature
is 295°C, flow rate of carrier gas 140 ml/min, catalyst
0.129 g, and pulse size 1.5 μl.

these linear plots and the activation energy from the Arrhenius plot:

$$k_2 = 1.13 \times 10^9 \exp(-28,200/RT) \ \text{sec}^{-1}$$

It has been suggested by many workers that the alkoxide group can be observed by infrared spectra and that it is an intermediate in the catalytic dehydration of ethanol. Consequently, the conclusion reached can be expressed as follows: Ethanol is adsorbed with alumina to form a relatively stable complex which is not desorbed and decomposes slowly to ethylene. Ether may be formed by interaction of ethanol vapor with the surface complex, i.e., the so-called Rideal mechanism. It can be assumed that the stable intermediate is analogous to alkoxide.

Ammoxidation of Toluene over a Vanadium Oxide Catalyst
    The reaction may be summarized as follows:

$$\text{C}_6\text{H}_5\text{CH}_3 + NH_3 + \frac{3}{2} O_2 \longrightarrow \text{C}_6\text{H}_5\text{CN} + 3H_2O \qquad (28)$$

    The following rate equation was suggested by a kinetic study in a flow reactor.

$$\frac{Q}{r} = \frac{1}{k_1 P_T} + \frac{1}{k_2 P_N} + \frac{1}{k_3 P_O} + \frac{1}{k_4} \qquad (29)$$

    Plots of the reciprocal of the overall reaction rate against the reciprocal of each partial pressure are found to be linear, as shown in Fig. 7. The rate equation is based on the following mechanism.

    (1) Toluene is irreversibly adsorbed on the catalyst surface with obstruction of the hydrogen of the methyl group by surface oxygen.

    (2) Adsorbed toluene reacts with ammonia to make benzonitrile, which can be desorbed easily.

    (3) Molecular oxygen is adsorbed on the reduced catalyst surface.

    (4) Then adsorbed oxygen is activated.

    Using the pulse reaction technique we obtained the number of active sites and the rate constant of each step, as explained in Section "Polystep Reactions." After injection of enough toluene, ammonia pulse reacts with adsorbed toluene to form benzonitrile; however, after repeated in-

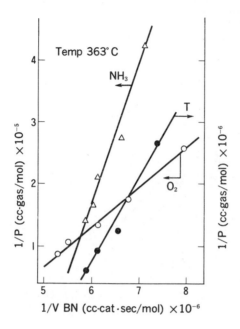

Fig. 7. Relation between rate of ammoxidation of toluene and partial pressures in the flow reaction.

jections of ammonia, benzonitrile is no longer formed, (Fig. 8). The activity of the catalyst is recovered by injection of oxygen. The rate constants and the concentrations of the active sites in the steps estimated are listed together with data from the flow system in Table I. These experimental results led us to the conclusion that ammoxidation of toluene seems to proceed by the four-step redox mechanism as proposed above (Fig. 9). All of the toluene adsorbed over the surface does not only to benzonitrile but also to carbon oxides. The benzonitrile formed was only 10 to 20% of the toluene adsorbed, and the residual toluene was still kept on the catalyst surface to be burnt by the oxygen pulse. It is probable that the vanadium oxide catalyst has at least two kinds of sites, one which promotes the formation of benzonitrile, and the other which does not, even though both sites

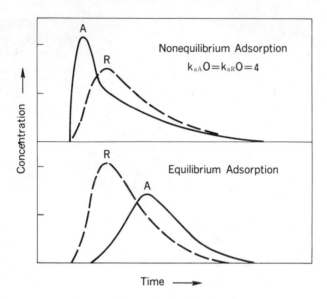

Fig. 8. Pulse reaction kinetics in step 2. Reaction between chemisorbed toluene and ammonia.

adsorb toluene. The fact that k_3 obtained in the pulse reaction is about ten times greater than that in the flow reaction can be explained on the basis of the existence of two kinds of active sites, because the sites not available for benzonitrile formation will be oxidized as well as the available sites.

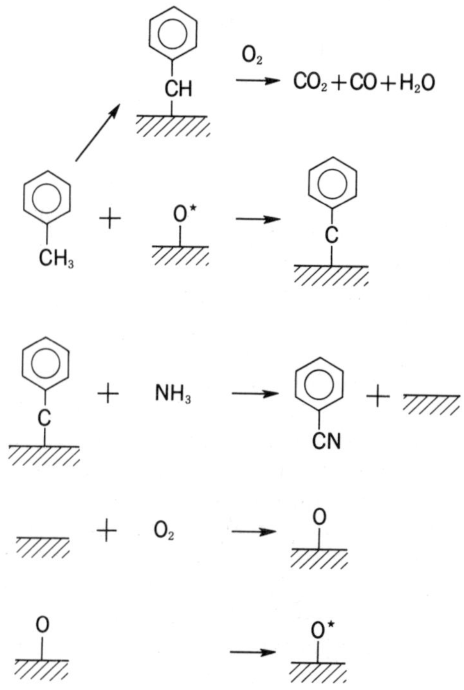

Fig. 9. Mechanism of ammoxidation of toluene.

Table I. Kinetic Constants in the Pulse and Flow Reactions.

Rate constant per active site (sec^{-1})	Flow reaction	Pulse reaction	Activation energy (kcal/mole)
k_1	2.02	2.35	10.6
k_2	0.340	0.340	11.7
k_3	0.0955	0.94	10.8
k_4	0.0606	-	11.7
No. of active sites (μmol/ml-cat)	6.70	2.11	19.7

Discussion

Y. Saito

The same Japanese computer IRA-5 as was used in the investigation on which Professor Slin'ko reported was used as an on-line computer to improve the abilities of the NMR spectrometer. That made it possible to examine rapid reactions with the help of this spectrometer. It was possible to analyze reactions at least one hundred times faster than usual ones. I used the IRA-5 computer with a magnetic core of 4 kw together with one magnetic drum of 8 kw. Sixteen sheets of rapidly scanned spectra were stored separately in the memory of both the drum and core through the channel of an AD converter. Each sheet was recorded separately in a digital form occupying the area of 1,240 words (each word had 296 points = 8 bits). After the reaction was completed DA conversion was used to give an electrical recorder spectrum. This technique was applied to the reaction of olefin oxymercuration and was reported last year at the Discussion Meeting on Catalysis in Japan.

S. L. Kiperman

In his report Professor Miyahara said that the method used by Horiuti did not require taking into account the mass action law. At the same time the concepts and equations of the theory of absolute reaction rates were used in the report. Thus, Professor Miyahara gives the values of reaction rates related with free activation energy as well as with other performance of an activated complex. However, as it is known, the theory of absolute reaction rates is directly related with the mass action law.

K. Miyahara

The theory of absolute reaction rates in the version proposed by Eyring is actually related with the mass action law. But we use a canonic ensemble method from which the expression also results which is analogous to the theory of absolute rates but without a mass action law.

G. Yablonski

I shall draw your attention to some inaccuracy in the analysis of the Christiansen stage sequence given in Professor Miyahara's report. From the equation of the rates of stationary reactions in Temkin form it can be easily shown that if the intermediate stage is irreversible the route rate does not seem to be independent upon the rate of stages following the irreversible one. However, the dependence does exist as it is necessary to know the concentrations of intermediates which are included into the rates of stages. Therefore, it is necessary on the graph which corresponds to a detail mechanism to change from peak to peak, i.e., from one intermediate to another, including into consideration those stages which follow the irreversible one as well.

Y. Moro-oka

I would like to refer to the relationship found in kinetic parameters.

In the case of a bimolecular reaction on a catalyst surface, i.e.,

$$A_{gas} \longrightarrow A_{ads}$$

$$B_{gas} \longrightarrow B_{ads}$$

$$A_{ads} + \quad B_{ads} \xrightarrow[slow]{k} C_{ads} \longrightarrow C_{gas}$$

the reaction rate is often expressed by the equation:

$$v = k' p_A^m p_B^n, \tag{1}$$

were k' is the apparent rate constant, p is the partial pressure, and m and n are the reaction orders.

Equation (1) is derived from the mass action law between the adsorbed species A_{ads} and B_{ads} on the catalyst surface,

$$v = k \, \theta_A \, \theta_B \tag{2}$$

were k is the rate constant of a surface reaction. θ_A and θ_B depend on the partial pressures according to the isotherm (Langmuir type, Temkin type, etc.).

For example, in case of Langmuir type

$$\theta_A = \frac{k_A p_A}{1 + k_A p_A} \simeq (k_A p_A)^m \tag{3}$$

The relationship $\theta_A \simeq (k_A p_A)^m$ can be also derived in case of other isotherm.

Thus, Eq. (2) can be written as follows:

$$v = k \, (k_A p_A)^m \cdot (k_B p_B)^n = (k k_A^m k_B^n) \, p_A^m p_B^n . \tag{4}$$

Then the apparent rate constant k' will be

$$k' = k \cdot k_A^m k_B^n = \exp\left(\frac{\Delta F}{RT}\right) \exp\left\{-\frac{m(\Delta F_{ads})_A}{RT}\right\} \exp\left\{-\frac{n(\Delta F_{ads})_B}{RT}\right\} \tag{5}$$

$$E_{obs} = E - m\Delta h_A - n\Delta h_B$$

$$\Delta S_{obs} = \Delta S + m \, (\Delta S_{ads})_A + (\Delta S_{ads})_B \tag{6}$$

where Δh_A and Δh_B is the heat of adsorption. Equation (6) was first derived by Schuit and Van Reijen. Since the observed activation energy E_{obs} and the observed activation entropy ΔS_{obs} are expressed by Equation (6), they do not correspond to any reaction step. Consequently it seems to be difficult to estimate the true catalytic activity from the observed activation energy or activation entropy. The method presented by Professor Miyahara makes it possible to estimate the energy state of an activated complex and does not depend on activation energy, activation entropy, and reaction orders. If the method is applicable to a complex reaction, it would be useful for the analysis of a hetero-geneous catalytic reaction.

M.I. Temkin

The conditions of the stationary state of a complex reaction can be formulated in different ways. The classical Bodenstein method involves the conditions of the type $d[\alpha_i]/dt = 0$, where $[\alpha_i]$ is the intermediate substance concentration. Another way (M.I. Temkin, DANSSSR (1963), 152, 156) involves using the equation

$$\sum_{p=1}^{p} \nu_s^{(p)} \gamma^{(p)} = \tau_s - \tau_{-s}$$

where $\nu_s^{(p)}$ is the stoichiometric number of Stage S over the route P, $\gamma^{(p)}$ is the reaction rate over route (p), summation

is performed over all basis routes, their number being
p; τ_s and τ_{-s} are the rates of Stage S in the direct and
reverse direction. The number of such equations is equal to
the number of stages; they are always sufficient to determine
unknown rates over routes and concentrations of intermediates. Sometimes one method is preferable and sometimes
another. From the equation given it is possible to derive
one other equation which often helps to obtain a kinetic
equation.

Professor Miyahara uses not a mass action law but a
general statistical-mechanical approach. As the mass action
law can be obtained as a result of the statistical-mechanical
approach, the results do not depend on the fact of whether
the mass action law is used or not.

The notes that follow on the reaction mechanisms discussed in the report, although not essential for its conclusions
as these reactions are used just for illustration, seem to be
not out of place.

The scheme of ammonia synthesis represented in the
report involves the stage of $N(a) + 3H(a) = NH_3$, i.e., adsorbed
nitrogen, is assumed to be hydrogenated by adsorbed hydrogen.

In studying the kinetics of ammonia synthesis near equilibrium it is impossible to get information on the mechanism
of adsorbed nitrogen hydrogenation, as the summary reaction
rate is determined by the rate of nitrogen adsorption.

However, if one uses very high volume rates of the
order $10^6 hr^{-1}$, it is possible to investigate the kinetics far
from equilibrium, as under these conditions two stages are
slow: nitrogen adsorption and addition of the first hydrogen
molecule. It permitted us to find out that the hydrogen molecule reacts out of a gas phase which collides with an adsorbed nitrogen molecule (N.M. Morozov, E.N. Shapatina, and
M.I. Temkin, Kinetika i Kataliz (1963), 4, 260, 565). The
similar conclusion was made later by Professor Tamaru on
the basis of the direct measurement of the hydrogenation
rate of nitrogen adsorbed on the ammonia synthesis catalyst.

While discussing ethylene oxidation on silver, Professor
Miyahara writes: "The presently adopted point of view on
the selectivity of ethylene oxide formation consists in that
the state of adsorbed oxygen causing the formation of ethylene
oxide differs from the state which results in full oxidation to
CO_2 and H_2O, the selectivity being dependent on their quantities."

Such a point of view was actually reported but it is not

universally adopted. At our laboratory we carefully studied
ethylene oxidation on silver, and in the corresponding articles
we proceed from the assumption that the same form of
adsorbed oxygen oxidizes ethylene both to ethylene oxide and
to CO_2 and H_2O (B.E. Ostrovski, N.V. Kulkova, M.S. Harson,
and M.I. Temkin, Kinetika i Kataliz (1964), 5, 469). In this
respect as well as in some others the mechanism proposed
by us is analogous to that proposed in the report. But there
are distinctions as well. In the report by Professor Miyahara
it is assumed that the intermediate substance at full oxidation
to CO_2 and H_2O is formaldehyde. In the work of S.Z. Roginski
and L.Ya. Margolis it was stated that with the help of ^{14}C as
labelled atoms that formaldehyde is not the intermediate
substance of the reaction discussed, therefore we supposed
another intermediate substance. The other distinction is
that the scheme of the reaction proposed in the report does
not take into account the adsorption of reaction products,
meanwhile such an account is necessary as all three products
- C_2H_4O, CO_2, and H_2O - slow down the reaction.

    I shall also note that the article cited which is dedicated
to ethylene oxidation includes an example of deducing kinetic
equations for the reaction with two routes with the help of
the above-mentioned general method which is different from
that described in the paper.

G.K. Boreskov
    First of all I would like to make a note in relation with
the speech by Professor M.I. Temkin. I would like to note
that the negative result obtained by the labelled-atom method
on the formation of a certain intermediate compound points
only to the fact that this compound does not pass into a gas
phase but does not exclude the formation and subsequent
conversions of this compound. Then I shall pass to the discussion of Golodets's report. Here a very important problem
of the catalytic reaction classification has been touched.
The problem is difficult, and it is important to discuss it.
I believe that the classification should be based on the
reaction mechanisms and interaction nature. Although he
speaks of these bases in the intoduction, the scheme proposed
is actually based on other approaches - the chemical composition of catalysts or the chemical nature of the whole process.
If the mechanism is not taken into account, the classification
is not fruitful. Thus, it is very important to base the classification of catalytic oxidation reactions on the differences in

oxygen-catalyst interaction. I do not insist on the oxidation-reduction mechanism. I admit the possibility of various mechanisms. But precisely this difference in mechanisms should provide the basis for the classification.

The use of differences in entropy values for the classification should be carefully treated, as the evaluation of activation energy presents difficulties and consequently a great mistake is possible in the determination of entropy. On the other hand entropy has low sensitivity to the mechanism changes. The same entropy can correspond to different reaction mechanisms. The introduction of a correction on the basis of a kinetic equation presents still more difficulties.

It seems to me that the report involves the not quite true statement that intermediates should be kept on the surface. One would think that on the contrary they should be more quickly eliminated from the surface.

In conclusion I would like to emphasize that the classification should be based on the difference in mechanisms. We do not always know the mechanism, but we should study it, i.e., without it a rational classification cannot be built and other basic problems of a heterogeneous catalysis cannot be tackled.

K. Miyahara

I think that the classification of catalysts given in the report by Roiter and Golodets is made by assuming the existence of a rate-determining step and considering its kinetic behaviour. However, such a case is rare. This is supported by the results of the recent experiments at our institute on stoichiometric numbers of heterogeneous catalysis. In many cases they are found not to be integers showing that the rates of constituent steps are comparable; hence the rate-determining step is absent.

I agree with the comments by Professor Temkin. However, I would like to emphasize that I tried to show the volume of information on the kinetic behaviour of constituent steps and respective intermediates. At present many researchers investigate the types and states of intermediates while they lack detailed, accurate measurements of variations of reactants in the course of a reaction. Therefore, I used some hypothetical or assumed sets of steps as examples of the application of my method. Up to the present the estimation of sets of steps has been outside my consideration, but now I am planning experiments to clarify the intermediates

and steps simultaneously with the determination of steady
rates.

V.S. Musykantov

One of the basic factors determining the catalytic properties of substances in reactions with the participation of
oxygen - the activity at complete oxidation and selectivity
at partial oxidation - is the strength of the oxygen-catalyst
bond. At present the role of this value at complete oxidation is clear enough: the catalytic activity is higher when
the bond strength is weaker.

The role of the oxygen bond strength at partial oxidation
is somewhat less elucidated. The only thing that can be
maintained convincingly is that for sufficiently high selectivity
the bond strength should be relatively high, as its decrease
will promote complete oxidation.

From our investigations of catalysts in the system V_2O_5 -
MoO_3 it follows that:

1)   the oxygen bond strength on the surface increases
while dissolving MoO_3 to V_2O_5 and reaches its maximum
value in the range of a saturated solution; and

2)   the selectivity at partial oxidation increases with
the increase of the oxygen bond strength. These conclusions
are quite the opposite of those made in the report by V.A.
Roiter and G.I. Golodets for the same system. The reason
for the discrepancies is connected with the methods of
determining the oxygen bond strength. With the sorption
method as a parameter of this value the adsorption heat
should be used, not the partial pressures of oxygen, as was
done in the report. The authors do not give temperature
dependences of oxygen equilibrium pressure, but from the
data reported by them earlier it follows that the maximum
partial pressure is also in agreement with the maximum
sorption heat, so that the bond strength in this case will be
maximum and not minimum, as is stated in the report. In
our work as a measure of the bond strength the values of
activation energy of oxygen heteroexchange were used that
were in close agreement with oxygen sorption heats on
catalysts.

Thus, the conclusion on the increase of selectivity while
increasing the oxygen bond strength of the catalysts investigated should be thought valid. It is possible that this conclusion is not of a universal character, as while comparing
various catalysts the change of the reaction mechanism may

have a pronounced effect on results, but in the series of one-type catalytic systems to which a number of solid solutions with different concentration of components can be attributed, the relation between the values under consideration can be clearly seen. The analogous relationship between oxygen bond strength and selectivity in relation to the number of reactions of partial oxidation is found by us for catalysts in the system Fe_2O_3 - Sb_2O_4.

S.A. Veniaminov

Professor Murakami made a very interesting communication on the use of a pulse reaction method to investigate heterogeneous catalytic reactions. In its usual modification the pulse reaction method is similar to a flowing one. As was shown by Professor Murakami, it can be achieved in the case when the pulse volume exceeds the catalyst volume by many fold.

At the Institute of Catalysis the procedure was developed in which the advantages of the pulse reaction method were combined with the advantages of a complete mixing reactor. The mixing is performed by intense vibration with the help of an electro-mechanical vibrator with the frequency of 50 c.p.s. and amplitude of 1-4 mm. The coincidence of the experimental results obtained in a pulse system with a vibro-liquefied catalyst layer with the results of a following-circulation method shows that in the pulse reaction method the conditions of complete mixing can be obtained. Such a procedure is very useful in studying catalysts with variable activity. We applied this procedure to the investigation of the mechanism of olefin oxidation in the presence of solid oxidation catalysts. The method made it possible to perform separate measurements of the rate of catalyst surface reduction by initial hydrocarbon and of the reduced oxygen surface oxidation.

A.E. Shylov

In connection with the report by Yu. I. Ermakov I would like to note the following: Low-valence derivatives of transition elements can be thought to be necessary for polymerization initiation only in the absence of alkyl aluminium derivatives and other metalloorganic compounds. In this case the structures of $M-CH_2CH_2-M$ type (M - transition metal) seem to be formed during olefin metal oxidation following which the introduction of a monomer over the M-C

bond is possible. In many works (including our experiments
on homogeneous catalysts of polymerization) it was soundly
established that polymerization can take place on metal
compounds with the highest valency (e.g., Ti (IV)) in the
case when alkylated derivatives of these metals are initially
formed, as a rule under the action of aluminium alkyls.
The polymerization itself, not including oxidation-reduction
steps, proceeds as the introduction over the M-C bond in the
case of high-valence M as well.

G.I. Golodets

On the basis of the data on equilibrium pressure of
adsorbed oxygen we showed that the stability of the oxygen
bond with the surface of V-Mo-oxidation catalysts passes its
maximum with the content of 30-50 mole % of MoO_3. This
conclusion is also supported by our data on initial rates of
surface reduction. In the comparison of oxygen bond stability
with catalytic properties, it must be taken into account that
we compared it with specific rates of selective oxidation and
not with selectivity, as was done in the work of V.S.
Muzykantov.

In connection with Professor Miyahara's comment we
shall note that the absence of a rate-determining step is not
a general case in the kinetics of heterogeneous catalytic
reactions. In many cases the rate-determining step can be
singled out as, for example, in the reaction of ammonia
synthesis on an iron catalyst proceeding not far from
equilibrium; in the processes of complete oxidation, decomposition with the evaluation of O_2, etc.

Like G.K. Boreskov we believe that a rational classification should be based on the resemblance of the mechanism of
catalytic reactions, as stated in our report. Unfortunately,
the detailed mechanism of most heterogeneous-catalytic
processes of oxidation has not been determined yet with sufficient reliability. The assumptions of the mechanism
cannot be considered as a suitable basis for classification.
It is much more reliable to proceed from the following
principle based on experimental data: if the reactions compared follow a similar mechanism then in an experiment the
resemblance of the catalytic activity series and optimum
catalysts should be observed. It is this principle that is used
as a basis for the classification proposed. The fruitfulness
of such an approach is proved by the fact that the analysis of
possible reasons determining the classification of catalytic

systems as certain types makes it possible to advance the representation of the mechanism as we have done in the example of incomplete oxidation of organic substances to acidic products. It should be emphasized that our classification is not a simple classification of reactions and not a classification by initial substances or products. It is the classification of the "reaction-catalyst" systems which takes into account the properties of all members of a catalytic process. The available data on the catalytic oxidation mechanism are in agreement with our classification. It concerns the reactions of complete and incomplete oxidation of olefins, alcohols, etc.

The comparison of the values of the activation true entropy is not the basis for a classification. It is one of the auxiliary criteria which can be used in search of the groups of one-type catalytic systems. Sometimes the values of ΔS^{*}_{true} have a very low sensitivity to the mechanism peculiarities, but in many cases such sensitivity clearly shows itself as, for example, at hydrogen oxidation at hydrocarbon complete oxidation on oxides and metals, etc. Therefore, there are no reasons to ignore those additional possibilities that are given by the analysis of the values of ΔS^{*}_{true}.

When we noted the necessity of stabilizing the product of incomplete oxidation on the surface, we kept in mind only the following: a good catalyst of incomplete oxidation should possess such properties that allow a surface complex to be formed on the surface which by its structure is near to the products of incomplete oxidation and can easily convert only to these products and not to the products of complete oxidation.

On the basis of the works of Academician G.K. Boreskov et al. (and some other works) it can be stated that on most oxidation catalysts of selective oxidation (at sufficiently high temperatures) a general redox scheme is being realized. Our classification is in agreement with these ideas. The nonstage ("associative") mechanism should be certainly taken into account in classification, but the presently available data are insufficient to allow it to be done accurately.

V.V. Gorodetski
One of the important questions of catalysis is the following: On which planes of metals are reactants adsorbed in the most active forms? This question cannot be answered with the investigations on polycrystalline samples. There are

experimental difficulties in proceeding on monocrystals.
We believe that the method using a field emission microscope
is the most acceptable, as the surface of the tip of a metal
under investigation represents all planes.

It is known that the adsorption of such gases as O_2 and
H_2 on metal films results in an increase of a work function
ϕ, but during the interaction of these gases on the surface
the decrease of ϕ is observed. The use of a field emission
microscope should result in a sharp increase of the brightness
of emission of those planes of the surface on which the interaction process takes place.

We obtained preliminary results from the investigation
of the interaction of hydrogen with oxygen adsorbed on
platinum surface. Platinum emitter was cleaned by pulse
heating up to 1750°K in ultrahigh vacuum of the order of 10^{-11}
torr. The change of a work function ϕ was calculated by a
Fowler-Nordheim equation at a constant emission current.

The adsorption of O_2 on Pt at 78°K (Fig. 1) results in an
increase of the work function by 1.40 eV. The heating of Pt
with adsorbed oxygen (curve 1) up to 600°K does not cause a
reasonable change of ϕ. The intorudtion of H_2 into the system
beginning at about 120°K (curve 2) results in a sharp decrease
of the work function of Pt, covered with adsorbed oxygen.

The observation of the change of field emission patterns
permits us to relate the decrease of ϕ with the process course
on different planes. Figure 2a represents the emission pattern of a clean surface of Pt emitter with orientation (III).
During the adsorption of O_2 up to saturation at T = 78°K and
$P_{O_2} = 5 \times 10^{-8}$ torr ($\Delta\phi$ = 1.40 eV), sharp specificity of adorp-
tion properties of different platinum planes was not observed
(Fig. 2b). At T = 78°K the interaction of $H_2(P_{H_2} = 5 \times 10^{-6}$ torr)
with adsorbed oxygen does not take place. At about 120°K in
the region around Plane (III) and on Planes (331) the interaction of hydrogen with adsorbed oxygen begins (Fig. 2c, $\Delta\phi$ =
1.35 eV). The increase of temperature from 120°K to about
180°K (Fig. 2d, T = 135°K, $\Delta\phi$ = 1.18 eV; and Fig. 2e, T =
150°K, $\Delta\phi$ = 1.06 eV) results in proceeding of the process on
Planes (102) $\Delta\phi$ being decreased to 0.85 eV (Fig. 2f). During
the interaction with H_2 the regions of these planes occupied
with adsorbed O_2 (dark regions) decrease in the dimensions
along the boundary of the adsorbed layer on the side of Plane
(III). It is possible that the reaction proceeds in the adsorbed
layer on the boundary of stationary O_{ads} (on Planes (102) and
adsorbed hydrogen (on the side of Plane (III)).

It is possible that the increased activity of oxygen adsorbed on Planes (331) and on the regions of the surface around Plane (III) is connected with a terrace structure of atom packing on these planes which seems to promote the transition of H atoms to oxygen atoms.

Y. Murakami

The pulse reaction method is reasonably simple experimentally, but the kinetic analysis of its results is often complicated and needs much mathematical analysis. I would like to emphasize that the pulse reaction method offers potential advantages. One of them is that we can concentrate our attention on a possible intermediate complex. It is very important, as surface adsorbates found by IRS, EPR, etc., methods are not necessarily true intermediate products, as Professor Tamaru pointed out. Another advantage is that this method permits us to study more exhaustively initial rates, poisoning effects, catalyst deactivation and other relaxation phenomena which are often useful in clarifying the mechanism of a catalytic reaction.

Activation of Molecular Nitrogen and Saturated Hydrocarbons by Transition Metal Complexes

A. E. SHILOV

Institute of Chemical Physics, Academy of Sciences of the USSR, Moscow

INTRODUCTION

Molecular nitrogen and paraffins are known to be chemically inert and usually undergo chemical reactions either at high temperatures or in the presence of highly reactive species.

The use of transition metal complexes allows us, in principle, to activate nitrogen and saturated hydrocarbons to enhance reactivity in mild conditions. In the case of N_2 the activation may proceed through molecular complexes:

$$M + N_2 \rightleftarrows MN_2, \qquad M + MN_2 \rightleftarrows MN_2M$$

The formation of mono- and binuclear complexes is due both to the donation of nitrogen electrons to free orbitals of the metal as well as to the back donation of d-electrons of the metal to empty antibonding π^* orbitals of nitrogen. All the known complexes have a linear structure.

For saturated hydrocarbons, oxidative addition to the metal may be expected:

$$M + RH \rightleftarrows M\begin{smallmatrix} R \\ \\ H \end{smallmatrix}$$

This is similar to the activation of other saturated molecules (H_2, HCl, CH_3I etc.). Though the activation of N_2 is different from the activation of C-H-containing compounds, there is certain similarity between these reactions. In oxidative addition, both empty and filled d-orbitals of the metal take part in formation of the new bonds. Thus in both cases (N_2 and C-H) low-valency metal compounds have to be used for the reaction.

Interaction proceeds in both cases according to the scheme "soft acid - soft base." The activation of molecular nitrogen and hydrocarbons may be expected in water solutions since a "hard base" — water—may be less reactive towards low valency metal compounds than towards nitrogen or hydrocarbons.

Activation and Fixation of Molecular Nitrogen

This work was started eight years ago. Like all our colleagues, we were inspired by the existence of biological N_2 fixation to ammonia at room temperature and atmospheric pressure and in the presence of water.

Soon we came to the conclusion which looks quite obvious now —that the first step in the activation of molecular nitrogen must be the formation of the complex with back donation of electrons to the nitrogen molecule as an important part of bond formation.

In 1965, Allen and Senoff prepared the first complex of molecular nitrogen with a transition metal compound.[2]

Strictly speaking, this well-known complex, $RuN_2(NH_3)_5$ X_2 should be regarded as the first inorganic diazocompound, since it was prepared not from molecular nitrogen but by reduction of $RuCl_3$ by N_2H_4.

However, we found earlier[1] that organic diazocompounds can be formed from N_2 and carbenes. Hence there was no further doubt that inorganic diazocompounds are in fact molecular nitrogen derivatives and can be prepared directly from N_2.

Indeed, we have found that N_2 forms a complex with bivalent ruthenium in tetrahydrofurane solution by reduction of $RuCl_3$ or $RuOHCl_3$ by amalgamated Zn.[3]

It was found later that N_2 could be complexed in quite "biological" conditions in the presence of H_2O and O_2 with a mild reducing agent such as Na_2S_2, Na_2S, or $TiCl_3$. However, we were unable to produce any reaction of the complexed nitrogen except for the evolution of N_2 into the gas phase.

Many stable complexes of N_2 were prepared by different authors, including binuclear complexes with two metal atoms. However for very long time nobody was to be able to reduce N_2 in a coordination sphere (see ref. 4).

It was necessary therefore, to realize the reason for this inertness, and so we made some thermodynamic calculations for consecutive breakdown of the triple bond.[5]

The energy of the first bond cleavage is so great—about 130 kcal/mole—that both one and two electron reactions with breakdown of this bond are usually endothermic, even for very reactive species such as atomic hydrogen:

$$H + N_2 \longrightarrow N_2H - 21 \text{ kcal}$$

$$H_2 + N_2 \longrightarrow N_2H_2 - 49 \text{ cal}$$

Similar reactions of acetylene are strongly exothermic. A large positive value of ΔH_f for N_2H_2 is the reason why hydrogenation of N_2 with intermediate N_2H_2 formation is impossible under the usual conditions.

Of course we can choose a reducing agent with an increased reduction potential to make the first two-electron steps possible, but in this case the reaction of nitride, formed at the end of the reduction, with H_2 and similar reducing agents is too endothermic to perform a catalytic cycle. We suggested a four-electron mechanism[4,5] to avoid a two-electron reduction step. We know that the binuclear complex is not a simple derivative of diimide since it is stabilized by $d_\pi - p_\pi$ interaction and N_2H_2 is not formed when water or acid is added. It is known that the ions forming the complex are not necessarily strong reducing agents.

In the subsequent formation of the hydrazine derivative, eash of M's give one electron, two more electrons being provided by a two-electron reducing agent present in the solution. The second π-bond in N_2 is considerably weaker than the first one, i.e., a much less strong reducing agent is sufficient for four-electron reduction of N_2 as compared with two-electron reduction.

It is of importance that the binuclear nitrogen complex allows the mechanism for formation of the hydrazine derivative.

Some interesting conclusions follow from this approach, particularly for reducing agents working in water and other protonic media. It may be shown that the catalyst M must be very specific.

Complex M will be catalytically active only inside a small range of M reduction potentials and M-N bond energies. The reason for this may be explained in the following way. If the reduction potential of M is too high, then the reducing agent present will not react with M to form a necessary valency state of reduced M, which would probably react with water instead of N_2 to form hydrogen.

If the reduction potential is too low, no reduction of N_2 to hydrazine will take place.

If M-N bond energy in the complex is too high the com-

plex will be too stable for the reduction to hydrazine derivative to proceed.   However, if the bond energy is too low,
then no complex will be formed at all.

These conclusions are valid only for protonic media,
but with certain modification, including the use of strong
reducing agents, they may be applied to aprotonic media.

The following predictions aries from this discussion.

1.   The reduction will proceed in the binuclear complex
to the hydrazine derivative, which is likely to be the next
intermediate.

2.   Intermediate complexes should preferably be
unstable.

3.   In water solution the reducing agent will be able to
reduce water to form hydrogen. However, the catalyst must
be very specific. Only a few metals, such as molybdenum in
nitrogenase in a particular oxidative stage might be catalytically active.

Now we have some experimental results which have confirmed these predications. From the above conclusions it
was natural to try first of all to observe unstable intermediate
complexes with N_2 at low temperatures in systems suitable for
the reduction of N_2. We have observed several complexes
which are thermally very unstable and exist only at low
temperatures.

One such complex was found in the system $CpTiCl_2$+
Grignard reagent[1,6] The complex formed bright blue solutions
and was in equilibrium with molecular nitrogen. Recently we
isolated the blue complex, analysed it, and found its formula
to be $(CP_2TiR)_2N_2$.

It turned out that hydrazine was the main product of the
reduction of the blue complex at -60 °C.[1,6,7]   Thereafter,
hydrazine was found to be the reaction product in several
systems.   Let us now consider the system containing
$FeCl_3$, i-PrMgCl + Ph_3P. A binuclear iron complex with N_2
was isolated and its formula according to chemical analysis
is $H[FeR(Ph_3P)_3]_2N_2$. In the i.r. spectrum of the complex
there is a band at 1760 cm^{-1} which corresponds to NN vibration, since it shifted to 1700 cm^{-1} when $^{14}N_2$ was substituted by
$^{15}N_2$.

Here we have found that the reduction to hydrazine proceeds only in the presence of HCl. No hydrazine was formed
when the complex decomposed in the presence of methyl
alcohol or carbon monoxide. The following scheme is the

mechanism we propose for this reaction:

$$\text{Ln} \underset{\text{FeN=NfeLn}}{\overset{\text{H}}{|}} + \text{H}^+ \longrightarrow \text{Ln} \underset{\overset{|}{\text{H}}}{\overset{\text{H}}{|}} \text{Fe-N=N} \overset{+}{-}\text{Fe} \longrightarrow \text{LnFeNH-NHFeLn}$$

$$\text{H}^{\mp}_{\text{N}_2\text{H}_4}.$$

HCl protonates complexed nitrogen, thereby increasing the electron accepting properties of the N_2 molecule, which can now be reduced to a hydrazine derivative.

As I have already mentioned, from the data on the biological reduction of nitrogen it follows that we could expect molybdenum compounds to be active towards molecular nitrogen in aqueous solution in the presence of reducing agents stronger than hydrogen.

According to our predictions, the first intermediate should be hydrazine. We have found that hydrazine is indeed the product of reaction in the systems where Cr^{2+}, V^{2+} and Ti^{3+} were chosen as the reducing agents and molybdenum - compounds act as catalyst. Hydrazine is the main product, especially at low temperatures, though ammonia was also detected. The reaction proceeds only in alkaline solutions. Many aspects of these reactions show definite similarities to enzymatic nitrogen fixation, e.g., the ratio of affinities towards CO and N_2 turned out to be very close to that in biological processes. Furthermore, it is important that Mg^2 salts are efficient both in biological and model systems. Moreover the only metal which could effectively replace Mo is V, just as in biological systems. We could expect V^{2+} compounds, which are themselves sufficiently strong reducing agents, to act without any extra reducing agent. Indeed we have found[8,9] that this is so, and it is V^{2+} compounds that form the most active species in water solutions in the presence of sufficient quantities of Mg^{2+} ions and alkali. The reaction proceeds at room temperature and atmospheric pressure of nitrogen. Nitrogen can be reduced directly from the air, the yield of hydrazine being half as much as when pure nitrogen is used.

This is the mechanism we suggest for reduction of nitrogen in the system Ti^{2+} + magnesium compounds.

$$-\underset{\underset{\displaystyle\overset{\displaystyle OH}{|}}{\vdots}}{Mo^{III}}\cdots N\equiv N\cdots \underset{\overset{\displaystyle\vdots}{H-O}}{Mo^{III}} \qquad -Mo^{IV}-NH-NH-Mo^{IV}-$$

$$-\underset{\displaystyle O}{\overset{|}{Ti^{III}}}\diagdown \qquad \underset{\diagup O}{Ti^{III}} \longrightarrow \quad -\underset{\displaystyle O}{\overset{\overset{\displaystyle O}{\|}}{Ti^{IV}}}\diagdown \qquad \underset{\diagup O}{\overset{\overset{\displaystyle O}{\|}}{Ti^{IV}}}$$

$$\diagdown\!\!\!\! Mg\diagup \qquad\qquad\qquad \diagdown\!\!\!\! Mg\diagup$$

The maximum yield of hydrazine is reached when Mg to Ti ratio is 0.5, and that is why we suggest this structure. This mechanism is by no means certain, and has to be confirmed.

Activation of Saturated Hydrocarbons

This work was started about five years ago. Though the activation of paraffins in heterogeneous catalysis has been well known for many years there has been no data until very recently for homogeneous solutions of transition metal complexes. In 1968, J. Halpern[10] enunciated this problem as follows. "The development of successful approaches to the activation of carbon-hydrogen bonds, particularly in saturated hydrocarbons, remains to be achieved and presently constitutes one of the most important and challenging problems in this whole field."

In our attempt to find the systems activating the C-H bond, we used the analogy of C-H and H-H- bonds. It is known that H_2 may be activated in solutions of transition metal complexes via oxidative addition, forming dihydrides which may dissociate in protonic media as acids. Since the M-C bond as a whole is weaker than the M-H bond, we focused our attention on the heavy metals of group VIII, i.e., Pt, Ir, and Os, which form stable alkyl derivatives. Alkylhydrides formed in oxidative addition reactions of RH with transition metal compounds in protonic media may be expected to interchange their protons with the protons of the solvent according to the scheme:

$$M + RH \rightleftharpoons M \underset{\diagdown H}{\overset{\diagup R}{}} \rightleftharpoons MR + H^+$$

Deuterium exchange with D_2O would be an indicator of oxidative addition of RH to M.

Indeed, we have found that when heating methane, ethane

and other paraffins with solutions of chlorides of Pt(II) and
Pt(IV) in heavy water, there is H-D exchange of D_2O with the
RH molecule.[11]

The reaction is inhibited by chloride ions which reduce
the metal compounds.

The kinetics of deuterium exchange for the case of ethane
have been studied to elucidate the mechanism.[12] Polydeuteroethanes were found to be formed, together with C_2H_5D in the
initial stages of the reaction, and their amount increases
simultaneously with that of monodeutoroethane. This indicates that more than one D atom is incorporated into the
ethane molecule during a reaction. The average number of
D atoms in the exchanged molecule was found to be 1.7. In
the presence of Pt(IV) compound, the kinetic curve has an
autocatalytic shape, indicating the necessity of compounds
other than Pt(IV) for catalytic activity. Spectroscopic
investigation has shown that during the induction period Pt(II)
complexes are formed which catalyse the exchange. Pt(IV)
compounds such as Pt(o) complexes have been found to be
inactive. The deuterium exchange proceeds with noticeable
rate in water solution but is 30 times faster in 50% acetic
acid (for the same ethane concentration in solution).

Thus the acetic acid molecule probably forms a weak
chelating ligand at Pt(II) atom and it is this which produces
the accelerating effect in the exchange reaction. The
dependence of the rate on the Cl^- and Pt (II) concentrations
is in agreement with the idea that solvated $PtCl_2$ is the most
active particle in the catalysis of the exchange.

Transition metal complexes in the reaction with hydrocarbons may possess stronger electron-accepting properties
than when they react with H_2 or halogenalkyls. Thus the
absence of Pt(O) activity to hydrocarbons is probably due to
its weak electron-accepting properties, while Pt(IV) is unable
to react in the oxidative addition of the C-H bond.

In agreement with the idea of electron-accepting properties of $PtCl_2$ for CH bond activation is the different
activity of different platinum chloride compounds, which are
in descending order of activity:

$$PtCl_2 > PtCl_3^- > PtCl_4^{2-}$$

and also

$$Cl > Br > I \quad \text{for } PtX_2$$

In connection with this, it is interesting to note that molecular hydrogen is activated by bases more easily than by acids, while for methane the converse is true.

Formation of C_2H_5D may be explained by the following sequence of reactions:

$$PtCl_2 + C_2H_6 \xrightarrow{-H^+} Cl_2PtC_2H_5^- \xrightarrow{+D^+} PtCl_2 + C_2H_5D$$

Independence of the ionic strength and noticeable isotope effect shows that the first stage is the rate-controlling step. The ratio of deuterium exchange for CH_4, C_2H_6, CH_3CO_2H (1:5:0.5) is in agreement with the electrophilic character of the oxidative addition for hydrocarbons. Formation of polydeuteroethanes might by explained by considering the possibility of the formation of intermediate complexes with ethylene:

$$Cl_2PtC_2H_5^- \xrightarrow{-H^+} Cl_2Pt \overset{CH_2^{2-}}{..CH_2} \xrightarrow{+D^+} Cl_2PtC_2H_4D^- \xrightarrow{+D^+}$$

$$C_2H_4D_2 + PtCl_2$$

Thus the formation of alkyl derivatives of a metal is the decisive step in the activation of saturated hydrocarbons.

References

1)   Yu. G. Borodko, A. E. Shilov, and A. A. Shteinman, Dokl. Akad. Nauk SSSR, 168, 581 (1966).

2)   A. D. Allen and C. V. Senoff, Chem. Commun, 621 (1965).

3)   A. E. Shilov, A. K. Shilova, and Yu. G. Borodko, Kinetika i Kataliz, 7, 768 (1966).

4)   Yu. G. Borodko and A. E. Shilov, Uspekhi Khim, 38, 761 (1969).

5)   G. I. Likhtenshtein and A. E. Shilov, Zh. Fiz, Khim., 44, 849 (1970).

6)   A. E. Shilov, A. K. Shilova, and E. F. Kvashina, Kinetika i Kataliz, 10, 1400 (1969).

7)   A. E. Shilov and A. K. Shilova, Zh. Fiz. Khim, 44, 288 (1970).

8)   N. T. Denisov, V. F. Shuvalov, N. I. Shuvalova, A. K. Shilova, and A. E. Shilov, Kinetika i Kataliz, 11, 813 (1970).

9)   A. E. Shilov, N. T. Denisov, O. N. Efimov, V. F. Shivalov,

N. I. Shuvalova, and A. E. Shilova, Nature, 231   (1971).

10)   J. Halpern, Trans. Farad. Soc. , 5   (1968).

11)   N. F. Goldshleger, M. B. Tjabin, A. E. Shilov, and A. A Shteinman, Zh. Fiz. Khim. , 43, 2174 (1969).

12)   M. B. Tjabin, A. E. Shilov, and A. A. Shteinman, Dokl. Acad Nauk SSSR, 198, 38 (1971).

Oxidation of Olefins by Mercuric Ion

Y. SAITO

University of Tokyo, Tokyo

INTRODUCTION

The oxidation of olefins, catalyzed by palladium (II) ion in aqueous solution to yield carbonyl compounds is well known as the Wacker process[1]. From the viewpoint of catalysis theory, the oxidation steps of olefins with palladium (II) ion in aqueous solution are more interesting than the catalytic reoxidation process of palladium (O) with oxygen. The reaction scheme, proposed by Henry[2] has now been widely accepted. A stable π-complex between the olefin and palladium (II) ion is slowly converted into β-hydroxyalkylpalladium (II) σ-complex, followed by the redox decomposition of the σ-complex as a fast step, with two electrons transferred simultaneously[3]. The mechanism of a double electron transfer is also postulated for thallium (III) oxidation of olefins in aqueous solution, where saturated carbonyl compounds, which are also produced in the palladium (II) oxidation, together with glycols[4]. The step of the π-complex formation between olefin and thallium (III) ion is assumed to be rate-determining, and it is succeeded by the conversion into σ-complex and its redox decomposition[5].

The redox decomposition of the σ-complex of these metal ions proceeds so fast that the elucidation of their reaction mechanisms is not easy, although the intramolecular hydrogen transfer has been demonstrated by Smidt[1] for the palladium reaction. This aspect of the olefin oxidation process by mercury (II) ion in aqueous solution should be worth investigating, since hydroxymercurated olefins, obtained as stable intermediates by reacting olefins with mercury (II) ion in aqueous solution[6] decompose into saturated carbonyl compounds under certain reaction conditions[7].

$$RHC=CHR' + Hg^{2+} + H_2O \longrightarrow RHC(OH)CHR'Hg^+ + H^+ \qquad (1)$$

$$RHC(OH)CHR'Hg^+ \longrightarrow RCOCH_2R' + Hg(0) + H^+ \qquad (2)$$

241

For example, acetone and methylethylketone are obtained from propylene and from either 1-butene, cis-2-butene or
trans-2-butene, respectively. These same compounds are
common to the reaction products of palladium (II) oxidation,
although unsaturated carbonyl compounds are formed with
different reaction conditions in the two types of reaction.[8,9]

The rate equation for the redox decomposition reaction,
shown in equation (2), was obtained as monomolecular with
respect to the mercurial, without depending on free
mercury (II) ion. The mechanism of the double electron
transfer between carbon and mercury (II) was therefore
strongly suggested.[10] However, the reaction rate of metallic
mercury with mercuric ion to give mercurous ion is very
fast and the equilibrium favors the mercurous side.[11]

$$Hg\,(0) + Hg^{2+} \rightleftharpoons Hg_2^{2+} \tag{3}$$

Indeed, mercurous ion was found in the aqueous solution by
Laser Raman spectroscopy after the reaction was over.[12]

The reaction mechanism to form β-hydroxy-sec-butyl
carbonium ion as a reaction intermediate as the result of
simple heterolysis of the C-Hg (II) bond is unfavorable, since
the reaction rate of hydroxymercurated cis-2-butene was
1.30 times faster than that of trans-2-butene, both yielding
methylethylketone.[12]

$$CH_3CH\,(OH)\,CHCH_3Hg^+ \longrightarrow CH_3CH\,(OH)\,CHCH_3^+ + Hg\,(O)$$

$$\longrightarrow CH_3COCH_2CH_3 + Hg\,(0) + H^+ \quad \text{(unfavorable)}$$

By means of time-sequential NMR spectroscopy monitored
in situ, together with the confirmation by mass spectroscopic
analysis for the reaction products, intramolecular hydrogen
transfer was found to take place during the reaction of redox
decomposition of hydroxymercurated cis-2-butene in D_2O
medium.[13]

$$CH_3CH\,(OD)\,CHCH_3Hg^+ \longrightarrow CH_3COCH_2CH_3 + Hg\,(0) + D^+ \tag{4}$$

In the present study, 2-deutero-trans-2-butene and 2-
deutero-cis-2-butene have been prepared and used to confirm
the observation stated above. Taking into consideration the
olefin dependence on rates of reaction, the reaction mechanism of the redox decomposition of hydroxymercurated

olefins is discussed. The reaction profile of olefin oxidations by other metal ions are also compared.

EXPERIMENTAL

Materials

Deuterated 2-butenes were prepared from 2-bromo-2-butenes, which were treated with metallic lithium in diethyl-ether at room temperature, and subsequently with deuterium oxide at 0°C. Two kinds of 2-bromo-2-butenes were obtained stereo-specifically by dehydrobromination of brominated trans- and cis-2-butenes. Normal gaseous olefins (ethylene, propylene, 1-butene, trans-2-butene and cis-2-butene) were all made by Takachiho Kagaku Kogyo Co., Ltd.. The absence of volatile impurities were confirmed in each case by gas chromatography using a 2-m β, β'-oxydipropionitrile column. Other chemical reagents were all of G.R. grade, made by Tokyo Kasei Kogyo Co., Ltd., and were used without further purification, except for diethylether, which was ordinarily refluxed with sodium after pretreatment with and were idendified on the basis of their NMR characteristics. The isomers were certainly formed by hydroxymercuration of 2-deutero-trans-2-butene. The NMR peaks due to these deuterated isomers decreased and those due to the reaction product increased with the elapse of time. A triplet pattern at $\tau 9.13$ from the hydrogen of the 4th position of the product methylethlketone, $CH_3CH_2COCH_3$, obtained from the normal mercurial as shown in Fig. 2, was in marked contrast to the doublet pattern at $\tau 9.15$ due to the corresponding hydrogen of the product, obtained from the deuterated one. It was identified as $CH_3CHDCOCH_3$ on the ground of NMR chemical shifts and coupling constants. The peak widths of the deuterated species were broadened by the nuclear spin-spin coupling between H and D, as far as the deuterium was attached at either the vicinal or the geminal position of the proton.

As the results of stereo-specific nature of hydroxymercuration of olefins, erythro and threo isomers were obtained from trans-2-butene and cis-2-butene, respectively,[14] which were clearly distinguishable from each other in the NMR spectra. The time-sequential NMR spectra for the redox decomposition of hydroxymercurated 2-deutero-cis-2-butene were obtained similarly. Similarly to trans-2-butene, the peaks of the two kinds of hydroxycalcium chloride in order to remove water.

244

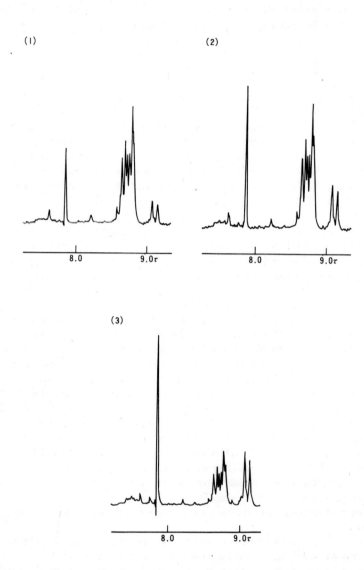

Fig. 1. The time-sequential NMR spectra for the redox decomposition of hydroxymercurated 2-deutero-trans-2-butene in aqueous solution. These spectra were taken after (1) 7 min. 55 sec., (2) 27 min. 01 sec. and (3) 83 min. 33 sec. from the moment of raising temperature.

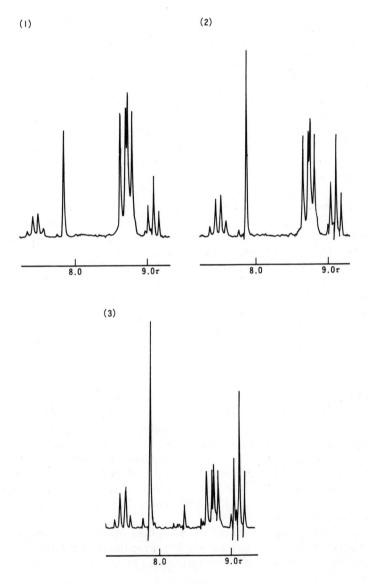

Fig. 2. The time-sequential NMR spectra for the redox decomposition of hydroxymercurated normal trans-2-butene in aqueous solution. These spectra were taken after (1) 14 min. 07 sec., (2) 35 min. 56 sec. and (3) 79 min. 13 sec. from the moment of raising temperature.

Oxidation of Olefins by Mercuric Perchlorate or Nitrate in
Aqueous Solution
      Deuterated 2-butene was transferred via a vacuum line
into a solution consisting of 3.00 M mercuric perchlorate
persisted at 0 °C until all mercuric ions were consumed by
hydroxymercuration. The 3.00 M solution of hydroxymercurated olefin was diluted with the same amount of 3.00
M mercuric perchlorate aqueous solution. Other olefins
were introduced at 0 °C into a solution consisting of 2.15 M
of mercuric nitrate and 1.10 M of nitric acid. A part of the
solution was transferred into an NMR sample tube and kept
in a dry ice-ethanol trap. The reaction was started by
inserting the tube into the probe of NMR spectrometer, by
which the solution temperature was raised up to the probe
temperature, regulated by blowing cooled nitrogen gas and
calibrated by the relative chemical shifts of 1,3-propanediol.
The reaction was recorded by taking sequential NMR spectra.
JEOL C-60 and PS-100 NMR spectrometers, made by Japan
Electron Optics Laboratory Co., Ltd., were used. A RMU-
S Mass spectrometer, made by Hitachi Co., Ltd., was used
to analyze the deuterium contents of 2-deutero-2-butenes and
methylethylketone.

RESULTS

The Redox Decomposition of Hydroxymercurated Deuterated
2-Butenes
      The redox decomposition reaction of hydroxymercurated
deutero-trans-2-butene in aqueous solution was monitored in
situ by NMR spectroscopy, as shown in Fig. 1. A set of
isomers, $CH_3CH (OH) CDCH_3Hg^+$ and $CH_3 CD (OH) CHCH_3Hg^+$,
was found in the spectra, mercurated 2-deutero-cis-2-butene,
$CH_3CH (OH) CDCH_3Hg^+$ and $CH_3CD (OH) CHCH_3Hg^+$, decreased
with the elapse of time, while the peaks due to the product
$CH_3CHDCOCH_3$ increased.
      The NMR parameters of the reactant mercurials and
the product ketones of the redox decomposition for normal
and deuterated 2-butenes are summarized in Tables I and II,
respectively.
      The results of mass spectroscopic analyses for 2-deutero-
trans-2-butene and its product are given in Table III. The
contributions due to d_1 in the products were exclusive.

Table I. NMR Parameters of Hydroxymercurated Normal and 2-Deutero-2-butenes.

Hydroxymercurated 2-butene	Chemical shift $(\tau)^a$							
	CH_3CHHg^+	CH_3CDHg^+	CH_3CHOH	CH_3CDOH	$CH_{-1}OHCHHg^+$	$CH_{-1}OHCDHg^+$	$CHOHCH_{-1}Hg^+$	$CDOHCH_{-1}Hg^+$
trans-2-butene (erythro isomer)	8.70 (d)		8.79 (d)		6.13 (o)		6.93 (o)	
	8.70 (d)	8.80 (s)	8.79 (d)	8.71 (s)		6.09 (q)		6.90 (q)
cis-2-butene (threo isomer)	8.62 (d)		8.81 (d)		6.14 (o)		6.99 (o)	
	8.62 (d)	8.82 (s)	8.81 (d)	8.63 (s)		6.10 (q)		6.95 (q)

Hydroxymercurated 2-butene	Coupling constant (Hz)		
	CH_3CHHg	CH_3CHOH	$CH_{-1}OHCH_{-1}Hg$
trans-2-butene (erythro isomer)	7.6	6.0	5.4
	7.5	6.0	
cis-2-butene (threo isomer)	7.6	6.0	4.0
	7.6	6.0	

a   The following abbreviations for the peak pattern are used: s:singlet, d:doublet, q:quartet, o:octet.

Table II. NMR Parameters of the Product Methylketone of the Redox Decomposition of
           Hydroxymercurated Normal and 2-Deutero-2-butenes.

	Chemical shift $(\tau)^a$			Coupling constant (Hz)
	$CH_3CH_2COCH_3$ $CH_3CHDCOCH_3$	$CH_3CH_2COCH_3$ $CH_3CH_2DCOCH_3$	$CH_3CH_2COCH_3$ $CH_3CHDCOCH_3$	$CH_3CH_2COCH_3$ $CH_3CH_2DCOCH_3$
Normal mercurial	9.13 (t)	7.50 (q)	7.87 (s)	7.3
Deuterated mercurial	9.15 (d)	7.49 (d)	7.87 (s)	7.2

a The following abbreviations for the peak pattern are used: s:singlet, d:doublet, t:triplet
   q:quartet.

Table III. Mass Spectroscopic Analyses of 2-Deutro-trans-2-butene and Its
            Product Methylketone.

Compound	d_0	d_1	d_2	d_3	d_4	d_5	d_6	d_7	d_8
2-Deutero-trans-2-butene	0.5	99.5	0.0	0.0	0.0	0.0	0.0	0.0	0.0
methylethylketone	0.8	99.2	0.0	0.0	0.0	0.0	0.0	0.0	0.0

Olefin Dependence of the Reaction Rates in the Redox Decomposition of β-Hydroxymercurated Olefins

The reaction rates of the redox decomposition of hydroxymercurated propylene, 1-butene, trans-2-butene, cis-2-
butene and cis-2-pentene were obtained from the time-
sequential NMR spectra pursued in situ. The reaction products were all ketones, namely, acetone from propylene,
methylethylketone from 1-butene, trans-2-butene, cis-2-butene,
and methylpropylketone with a small amount of diethylketone
from cis-2-pentene. The first order rate equations with respect to the hydroxymercurated olefins were confirmed in each
case by the fact that the three kinds of plots, obtained from
the peak decrease of the reactant mercurial versus the external reference, tetramethylsilane, from the peak increase
of the product ketone versus the reference, and from the
mercurial decrease versus the ketone increase without the
intermediary of the reference, were obtained in parallel.
From the coincidence of these slopes, which reflects stoichiometric nature of the reaction, it was ascertained that ketones
were the sole products with the reaction conditions adopted in
this study.

Acetone, produced from propylene, was stable in solution at a temperature as high as 60°C, but a consecutive reaction was distinctly observed above 40°C in the NMR spectra for methylethylketone produced from butenes. With regard to the redox decomposition of hydroxymercurated 2-butenes, which was performed below 30°C, this consecutive reaction of methylethylketone did not influence the reaction analysis. In the case of 1-butene, however, the redox decomposition was performed above 35°C because of its moderate reactivity, so that the product peaks had to be discarded in the analysis. For this reason the rate constants for hydroxymercurated 1-butene should be treated as being less accurate than the others. Regarding cis-2-butene, the reaction temperatures below 20°C prevented any consecutive reaction of the products. The NMR spectra also revealed that methylpropylketone and diethylketone were formed from $CH_3CH(OH)CHC_2H_5Hg^+$ and $C_2H_5CH(OH)CHCH_3Hg^+$, respectively, the former greatly predominating over the latter. The reaction rate was obtained, therefore, for the formation of methylpropylketone only. Hydroxymercurated ethylene was so stable that no change was observed in NMR spectra up to 70°C.

The rate constants of the redox decomposition of hydroxymercurated olefins in aqueous solution were obtained as a function of temperature, as shown in Fig. 3. The activation energies were summaried in Table IV, together with the ketone products.

DISCUSSION

Intramolecular Hydrogen Transfer in the Process of the Redox Decomposition of Hydroxymercurated 2-Butenes

It was previously ascertained by means of both NMR and mass spectroscopy that no deuterium was incorporated during the redox decomposition of hydroxymercurated cis-2-butene in D_2O medium.[13] In the present study, the deuterated 2-butenes were confirmed by mass spectroscopy to change into methylethylketone, with one deuterium replaced. The deuterium contents of all of them are exclusively d_1, as shown in Table III. The position of deuterium replacement in the starting materials of trans-2-butene and cis-2-butene is apparent not only from the viewpoint of the method of synthesis but also appeared in the NMR spectra, as a set of isomers of hydroxymercurated 2-butenes, identification of which is confirmed by the NMR characteristics given in

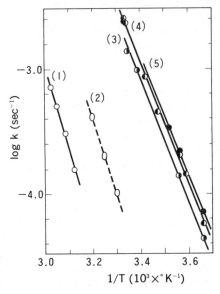

Fig. 3. The rate constants of the redox decomposition of hydroxymercurated olefins in aqueous solution as a function of temperature. Olefins were (1) propylene, (2) 1-butene, (3) trans-2-butene, (4) cis-2-butene and (5) cis-2-pentene.

Table IV. The Activation Energies and the Product Ketones of the Redox Decomposition Reactions of Hydroxymercurated Olefins in Aqueous Solution

Olefin	Activation energy (kcal mole⁻¹)	Product ketone
Propylene	29.7	acetone
1-butene	28	methylethylketone
trans-2-butene	23.0	methylethylketone
cis-2-butene	22.3	methylethylketone
cis-2-pentene	21.4	methylpropylketone

Table I. The position of deuterium replacement in the product methylethylketone is also confirmed by the NMR spectra shown in Fig. 1 as well as in Table II.

It is therefore evident that intramolecular hydrogen transfer takes place in the redox decomposition reaction of hydroxymercurated 2-butenes:

$$CH_3CH(OH)CDCH_3Hg^+ \longrightarrow CH_3COCHDCH_3 + Hg(0) + H^+ \qquad (5)$$

$$CH_3CD(OH)CHCH_3Hg^+ \longrightarrow CH_3COCHDCH_3 + Hg(0) + H^+ \qquad (6)$$

In the palladium (II) oxidation of olefins in aqueous solution, it is well known that intramolecular hydrogen transfer takes place at the step of the redox decomposition of β-hydroxyalkyl palladium (II) σ-complex, as no deuterium incorporation in the product CH_3CHO was observed for C_2H_4 in D_2O medium.[1] A four-centered mechanism was proposed for this concerted reaction on the grounds of the affinity between palladium (II) and hydride ion:

$$
\begin{array}{c}
\underset{\displaystyle H-\underset{\displaystyle Pd^+\,OH}{\overset{\displaystyle H\quad H}{C-C-H}}}{}
\longrightarrow
\left[\underset{\displaystyle Pd\text{---}H}{\overset{\displaystyle H\quad OH}{H-C-C-H}} \right]^+
\longrightarrow
\underset{\displaystyle H}{\overset{\displaystyle H\quad O}{H-C-C-H}} + Pd(0) + H^+
\end{array}
\qquad (7)
$$

Though no detailed information is available concerning the bonding between mercury (II) and hydride ion, an estimation of low affinity was made when the heterolytic splitting mechanism of molecular hydrogen with mercuric ion was proposed.[15] With respect to the stereochemistry of oxymetallation of olefins, a ligand insertion or a syn addition mechanism is widely accepted for oxypalladation,[3] whereas the oxymercuration of free olefins has been ascertained as an anti addition.[6] Thus the behaviors of palladium (II) and mercury (II) ion toward olefins or hydride ion are in sharp contrast in some respects.

As far as the reaction mechanism of the redox decomposition of hydroxymercurated olefins is concerned, a concerted mechanism for this oxidative rearrangement would be strongly suggested from the viewpoint of intramolecular hydrogen transfer, but it is still undetermined whether a syn mechanism, i.e. a four-centered one, is occurring, as was assumed in the case of palladium (II), or whether an anti

mechanism, i.e., a three-centered one, takes place.
According to a preliminary analysis of the kinetic deuterium
isotope effects for the redox decomposition of hydroxymercurated 2-butenes, the latter mechanism, shown in Eq.
(8), seems to be probable:

$$
\begin{array}{c}
\text{H} \ \text{OH} \\
| \ \ | \\
\text{CH}_3-\text{C}-\text{C}-\text{CH}_3 \\
/ \ \ | \\
\text{Hg}^+ \ \text{H}
\end{array}
\longrightarrow
\left[
\begin{array}{c}
\text{H} \ \ \ \text{H} \ \ \ \text{OH} \\
\backslash \ \ \ \ \ / \\
\text{C}-\text{C} \\
/ \ \ \ \ \ \backslash \\
\text{CH}_3 \ \text{Hg} \ \text{CH}_3
\end{array}
\right]^+
\longrightarrow
\begin{array}{c}
\text{H} \ \ \text{O} \\
| \ \ \| \\
\text{CH}_3-\text{C}-\text{C}-\text{CH}_3 \ +\text{Hg}(0)+\text{H}^+ \ \ (8) \\
| \\
\text{H}
\end{array}
$$

Olefin Dependence of the Reaction Rates and the Carbonium
Ion Character of the Activated Complex
       According to the standpoint of the linear free energy
relationships, the olefin dependence of the reaction rates for
the redox decomposition of hydroxymercurated olefins would
be correlated with the difference of the free energy of
formation of the reactant mercurials and the ketone products.
Since the equilibrium constants for hydroxymercurated olefin
in aqueous solution were recently reported on the basis of
the rate constants of both oxymercuration and deoxymercuration of a given olefin, obtained by a stopped flow technique,[16]
it is now possible to assume the values of the free energy of
formation of hydroxymercurated olefins by combining those
of gaseous olefins[17] with these equilibrium constants. The
free energy changes owing to dissolution of gaseous olefins
and ketones into aqueous solution are not taken into consideration intentionally, because this estimation is aimed at
examining the linear free energy relationships, and the
factors determining the solvation energy of olefins, ketones
and the activated complexes in aqueous solution, would not be
straightforward.
       The rate constants of the redox decomposition at 25.0°C,
given by interpolating or extrapolating the results at other
reaction temperatures, and by the free energies of formation
of the reactant mercurials estimated above and of gaseous
ketones,[7] together with the differences between them, are
summarized in Table V. Although the rate constants vary
more than 10^3 times among the reacting olefins, the free
energy differences vary only within the range of 2.2 kcal.mole^{-1}.
Direct application of the linear free energy relationships to
this reaction system seems rather difficult. The reactivity
difference between trans-2-butene and cis-2-butene might

Table V. The Rate Constants of the Redox Decomposition for Various Hydroxymercurated Olefins in Comparison with the Free Energy Difference of the Reaction and the Taft σ^* Parameters of the Substituents.

Olefin	Rate constant of redox decomposition[a]	Free energy of formation (kcal mole⁻¹)			Taft σ^* parameter of substituents[c]	
		Hydroxymercurated olefin	Carbonyl product	Difference[b] Difference	R	R
ethylene	–	6.6	-31.86	38.5	0.49	0.49
propylene	4.93×10^{-7}	5.27	-36.58	41.85	0.49	0.00
1-butene	4×10^{-6}	7.84	-34.91	42.75	0.49	-0.10
trans-2-butene	1.47×10^{-3}	7.63	-34.91	42.54	0.00	0.00
cis-2-butene	2.14×10^{-3}	9.14	-34.91	44.05	0.00	0.00
cis-2-pentene	2.16×10^{-3}	10.6[d]	-32.76	43.4	-0.10	0.00

a   The rate constant at 25.0°C were given.
b   The difference was given as ΔG_f^0 (reactant) - ΔG_f^0 (product).
c   The substituents were designated as RCH(OH)CHR'Hg⁺.
d   The equilibrium constant, estimated as 6×10^4 was used.

reflect the free energy difference of erythro and threo isomers
of the hydroxymercurated 2-butenes. But this would not be
the case, because the reactivity of 1-butene cannot be under-
stood from this standpoint. The difference should probably
be interpreted as the result of the concerted mechanism,
which consists of simultaneous intramolecular hydrogen
transfer and two-electron transfer during the redox decomposition.

It should be noted for the rate constants that hydroxymercurated terminal olefins, i.e. ethylene, propylene and
1-butene, are less reactive than internal olefins, i.e.,
trans-2-butene, cis-2-butene and cis-2-pentene, suggesting
that the activated complexes have a character similar to
some extent to carbonium ion. Indeed, some related reactions
such as oxymercuration of olefins and its deoxymercuration
exhibit carbonium ion character, since the log k for these
reactions are linearly correlated with Taft σ^*, with ρ^* of -3.3
for oxymercuration[18] and -2.77 for deoxymercuration.[19]
In Table V, the Taft σ^* parameters are also included for the
substituents at the two positions of the mercurial, RCH (OH)
CHR′Hg$^+$. Though data are scarce, a similar tendency of
carbonium ion character was deduced from the set of propylene, 2-butenes and cis-2-pentene of R = H, CH$_3$ and
C$_2$H$_5$, keeping R′ as CH$_3$. The dependence on the R′ substi-
tuent is also undecided, but a similar tendency seems to be
pointed out from the series of ethylene, propylene and 1-
butene. A certain extent of carbonium ion character would
be undeniable for the activated complex of the redox decomposition of hydroxymercurated olefins. This is also in contrast to the case of palladium (II) oxidation of olefins.

However, the most distinct difference between mercury
(II) and palladium (II) ions in the oxidation of olefins is found
in the profile of reaction sequences. The olefin π-complex
of mercury (II) ion has been assumed as the reaction intermediate of hydroxymercuration, which is far more rapid
than that of the redox decomposition of hydroxymercurated
olefin. Although the free energy of formation of the π-com-
plexed intermediate, is, of course, uncertain, it must be
higher than that of reactant olefin as well as of hydroxymercurated olefin. Conversely, the olefin π-complex of
palladium (II) ion is stable and formed as the pre-equilibrium
product, whereas β-hydroxyalkyl palladium (II) complex is
formed as the product of the rate-determining step. The
free energy of formation of this σ-complex of palladium (II)

must be high in contrast to the corresponding σ-complex of mercury (II), shown in Table V.

Both palladium (II) and mercury (II) ion are classified as soft acid, according to the HSAB principle,[20] but the properties of carbon, which is classified as soft base, are different from each other, as reflected in the free energies of formation of the olefin σ- and π-complexes, which would have their origin in the differences in the back-donating properties. The characteristics of soft metal ions, favoring σ-carbon as mercury (II) or rather π-carbon as palladium (II), are important for the theory of catalysis, as it exerts a deep influence on the reaction profile or the position of the rate determining step. In this respect, thallium (III) in olefin oxidation could be characterized as a soft metal ion as it favors σ-carbon less than mercury (II) and π-carbon far less than palladium (II).

As far as the differentiations due to the σ-soft and π-soft properties are taken into consideration, the HSAB principle seems to be useful for understanding chemical reactions, including catalysis.

Acknowledgements

The author is grateful to Professor Y. Yoneda for helpful discussion and encouragement and to Dr. M. Matsuo and Mr. S. Shinoda for collaboration in experiments.

References

1) J. Smidt, W. Hafner, R. Jira, R. Sieber, J. Sedlmeier, and A. Sabel, Angew. Chem. Intern. Ed. Engl., 1, 80 (1962).
2) P.M. Henry, J. Am. Chem. Soc., 86, 3246 (1964).
3) E.W. Stern, Catalysis Reviews, 1, 74 (1968); A. Aguilo, Advan. Organometal. Chem., 5, 321 (1967).
4) P.M. Henry, J. Am. Chem. Soc., 87, 990, 4423 (1965); 88, 1597 (1966).
5) P.M. Henry, Advan. Chem. Ser., 52, 126 (1968).
6) W. Kitching, Organometal. Chem. Rev., 3, 61 (1968).
7) Y. Saito and M. Matsuo, J. Organometal. Chem., 10, 524 (1967).
8) B.C. Fielding and J. Roberts, J. Chem. Soc., A, 1627 (1966).
9) J.C. Strini and J. Metzger, Bull. Soc. Chim. France, 3145, 3150 (1966).
10) M. Matsuo and Y. Saito, Bull. Chem. Soc. Japan, in press.

11) F.A. Cotton and G. Wilkinson, in Advanced Inorganic Chemistry, Interscience (1962), p.481.

12) M. Matsuo, PhD dissertation at the University of Tokyo (1971).

13) M. Matsuo and Y. Saito, submitted to publication.

14) M.M. Kreevoy, L.L. Schaleger, and J.C. Ware, Trans. Farad. Soc., 56, 2433 (1962).

15) J. Halpern, in Proc. 3rd Int. Cong. Catalysis, 1965, Vol. 1, p.146.

16) J.E. Byrd and J. Halpern, J. Am. Chem. Soc., 92, 6967 (1970).

17) D.R. Stull, E.F. Westrum, Jr., and G.C. Sinke, The Chemical Thermodynamics of Organic Compounds, John Wiley, 1969.

18) J. Halpern and H.B. Tinker, J. Am. Chem. Soc., 89, 6427 (1967).

19) L.L. Schaleger, M.A. Turner, T.C. Chamberlin, and M.M. Kreevoy, J. Org. Chem., 27, 3421 (1962).

20) R.G. Pearson, J. Am. Chem. Soc., 85, 3533 (1963); J. Chem. Educ., 45, 581, 643 (1968).

APPENDIX

Comparison of Metal Ion Oxidation of Olefin

The magnitudes of the reaction rate of olefin oxidation to yield saturated carbonyl compounds by Pd(II) ion in aqueous solution are in the order of ethylene, propylene, butene, etc. Olefin dependence on oxidation by Hg(II) ion giving the same products is the reverse, as shown in Fig. 3 and in Table IV. The higher the carbon number of olefins, the larger the reaction rates in the mercury case. Indeed the olefin dependence on the reaction rates is different between the two, but the reaction products are common and a concerted mechanism is postulated for both, including both intramolecular hydrogen transfer and two-electron transfer.

According to Henry,[5] olefin oxidation by Pd(II) ion proceeds along the following reaction path:

olefin ⟶ π-complex ↭ σ-complex ⟶ saturated carbonyl cmpds.

With regard to hydroxymercuration of olefins, π-complex intermediate (mercurinium ion) is usually assumed.[6] Since hydroxymercuration products are rather stable, the total reaction rate yielding ketone from olefin is limited by the redox decomposition step of σ-complex. The reaction path would be, therefore, represented for the mercury case as follows:

olefin ⟶ π-complex ⟶ σ-complex ↭ saturated carbonyl cmpds.

As for the Tl(III) ion oxidation of olefins, by which ethylene is changed into acetaldehyde, the rate-determining process was suggested to be the step of π-complex formation.[5]

olefin ↭ π-complex ⟶ σ-complex ⟶ saturated carbonyl compds.

Thus metal ion oxidations of olefins to yield saturated carbonyl compounds seem to proceed along the same reaction path for Pd(II), Hg(II) and Tl(III), the rate-determining steps being different from each other. The characteristics of these metal ions would be attributed to the relative stability of π-complex and σ-complex of metal ions, i.e., Pd(II) prefers π-complex owing to its back-donation ability. Hg(II) prefers σ-complex, as donation is prevailing. Tl(III) does not prefers neither, since both donation and back-donation would be weaker than for the other two. These characteristics in chemical bonding for the reaction intermediates would cause the differences in olefin dependence on rates between Pd(II) and Hg(II) oxidation.

(Because of the time limit for presentation, the last section, "Comparison of metal ion oxidations of olefin", was read at the seminar instead of "Olefin dependence of the reaction rates and the carbonium ion character of the activated complex.")

Optimization of Catalysts for Hydrogenation in Solutions

D. V. SOKOL'SKY and A. M. SOKOL'SKAYA

Institute of Organic Catalysis, Academy of Sciences of Kas. SSR, Alma-Ata

Today, after numerous investigations carried out by various schools of catalysis study, the following assertions may be considered proven.

1) The specific action of a catalyst depends only on its chemical composition (lattice structure, electronic constitution, chemical properties), i. e., the position of the element in the Periodic System.

A particularly detailed development of this assertion is given by G. K. Boreskov and his school, who found that the catalytic activity per unit of its real surface is little dependent (for the given reaction) on the catalyst preparation method. The frequently observed apparent exceptions to this rule may be explained by the effects of different impurities due to variation in the preparation method.

2) The activity of a catalyst mass unit (1 g) depends on its surface constitution, pore volume and other physical factors. Now it is possible to prepare catalysts of a predetermined size (metal-ceramics, zeolites etc.), such as has been achieved by P. A. Rebinder and his school.

The surface layer composition of the catalyst often differs from the composition of the volume phase.

3) The activity of a supported atomic phase is without doubt, being noted by N. I. Kobozev as long ago as 1939.

4) The most reliable method of investigating the catalyst surface layer properties is to study the directions of different reactions on a given catalyst and to measure the reactant and radical concentrations on the surface.

5) For further development of catalysis, a precise definition of A. A. Balandin's assertions (constituting his multiplet theory) concerning structural and energetic correspondence is essential.

The required structure and energy correspondences should not be considered apart from one another. Only in single cases do these two aspects act separately.

Our view is that special attention should be paid to the

existence of various forms of energy correspondence[1].
The following modifications, for instance, exist in the
case of catalytic hydrogenation.

a) Values of or any other reactant bond energies between
catalyst and hydrogen,

b) Electron work function,

c) Dissolved hydrogen (proton) work function,

d) Instability constant of the complex compound (in the
homogeneous catalysis), etc.

The energy correspondence (and hence the structural
correspondence) presents itself in different forms in radical,
ionic and electronic surface processes.

Apparently, in oxidation-reduction reactions, the surface
acidic-basic properties are determining for the structural
correspondence, i. e. , the molecule orientation.

The use of electrochemical techniques in the investigation
of catalysts and catalytic reaction mechanism allows us in
some cases to distinguish between the effects of various
factors on compound adsorption and its reactive capacity
(reactivity) in conditions of heterogeneous-catalytic
conversion.

Hydrogenation in solutions may essentially proceed in two
different ways[1,2] depending on the kind of catalyst and
unsaturated compound.

1. During the reaction, the catalyst potential remains
close to the hydrogen potential in the given solution.
Hydrogenation proceeds with the catalyst surface almost
completely covered with hydrogen. With these conditions
the reaction is first order in respect to unsaturated compound
and zero or fractional in respect to hydrogen. For materials
weakly displacing hydrogen from the catalyst surface (hexane,
substituted olefins etc.) this type of hydrogenation occurs
even at normal pressure and particularly in alkaline solutions.
The hydrogenation reaction rate of such materials decreases
with pH increase.

Of materials readily and quickly taking hydrogen off the
catalyst surface (acetylene, its homologus and derivatives,
nitro-compounds and compounds with conjugated bonds), a
zero order of reaction is characteristic for hydrogen when
the rate of its reproduction on the catalyst surface is greater
than the take-off rate. This occurs at a pressure of 40-60
atmospheres. The apparent activation energy depends on the
unsaturated compound structure and fluctuates within the
range of 4-9 kcal/mole for acetylenic and ethylenic

hydrocarbons.

2. During the reaction, the potential shifts towards the anodic side (relative to the hydrogen potential) by a value, practically excluding the presence of hydrogen on the catalyst surface ($\Delta E > 250$ mV). In this case zero order of reaction is observed with unsaturated compounds and first order with hydrogen. This type is realized in hydrogenation of materials that readily and completely remove hydrogen from the catalyst surface (quinone, nitrobenzene, vinylacetylene etc.). The activation energy of the reactions is characteristic of the hydrogen atomization process and is practically the same in hydrogenation of any compound, reaching 12 - 14 kcal/mole.

In some cases the first step of the reactions is the electron transfer from the catalyst surface to the unsaturated compound, forming a negatively charged radical (e.g., hydrogenation of oxygen, quinone, nitrobenzene, acetylene over palladium). The hydrogenation reaction rate of such compounds does not depend on the solution pH.

Intermediate cases are possible, when the extent to which the catalyst surface is covered with hydrogen is a function of the hydrogenated compound structure. Therefore, the maximum hydrogenation reaction rate is observed at a concentration of reactants on the surface (hydrogen or an unsaturated compound) close to the stoichiometric concentrations. The hydrogenation reaction rate of such compounds can pass through a maximum or a minimum with increase in the pH value. Especially characteristic are the experimental dependence forms of the multifunctional compounds. By a series of examples it was shown that in acidic solutions the organic molecules on platinum, palladium and nickel surfaces undergo considerable decomposition even at relatively low temperatures (20-40 °C).

The linear dependence of the catalyst potential shift on the degree of hydrogen on the surface (up to complete hydrogen take-off corresponds to the logarithmic dependence on the hydrogen pressure. This corresponds to a linear dependence of the hydrogen-reaction-activation energy in current-conducting solutions on the potential shift. The activation energy of the hydrogenation reaction can be calculated as follows:

$$A = \Delta E \times 23 \pm B \tag{1}$$

where A is the activation energy in kcal/mole, ΔE the potential shift in volts during the reaction, and B a constant value dependent on the kind of catalyst (within the range of 4-7 kcal/mole). Minus sign appears after $\Delta E > 0.4v$.

When calculating the true activation energy on the basis of the constant values of hydrogenation rates in accordance with the Arrhenius formula, we must be sure that the catalyst potential was the same at different temperature (ΔE). If it was not, then the difference in the potential values permits a direct evaluation of the pre-exponential term.

In most cases the hydrogenation rate decreases during the reaction due to adsorption of the reaction product or products. Within a very wide range of hydrogen pressures, the hydrogenation reaction rate obeys the equation:

$$v = K \, P_{H_2} \frac{b_1 \, c_1}{b_1 c_1 + b_2 c_2 + b_3 c_3} \exp\{-(\Delta E \times 23 \pm B)\} \qquad (2)$$

where b_1, b_2 and b_3 are the adsorption coefficients of the unsaturated compound, the reaction product and the solvent, and c_1, c_2 and c_3 are the corresponding concentrations.

During hydrogenation in hydrocarbons or at a strong adsorption of the unsaturated compound, the solvent adsorption may be neglected. However, the solvent can influence the reproduction rate of the active hydrogen and its bond energy with the surface of the catalyst.

During adsorption of the unsaturated compound, preliminary displacement of the solvent molecules from the catalyst surface must be carried out. In spite of this, adsorption seems to remain isothermic. In steady-state conditions the catalyst surface may be almost completely free of solvent molecules.

With the completion of these studies, we are now in a position to select an optimal catalyst for the hydrogenation reaction. It is possible to predict the required optimal heat of adsorption and the energy of the bond between the reactant atoms and the catalyst surface.

However, how to prepare such a catalyst is quite another question. There are difficulties in the preparation of a catalyst of given dispersivity and porosity. Naturally, one of the main objects is the selection of an optimal catalyst and conditions of hydrogenation in solutions.

Figure 1 shows the results of an investigation into the hydrogenation reaction of benzaldehyde in ethanol at 30 °C

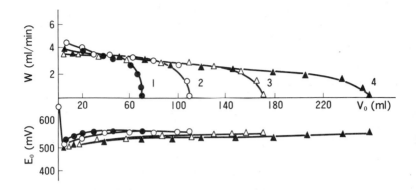

Fig. 1.   Hydrogenation of various amounts of benzaldehyde
over a 0.5 % Ni-Pd catalyst (0.5 g) in ethanol at 30°C;
Curve 1,   0.3 ml;   2,   0.5 ml;   3,   0.8 ml;   4,   1.2 ml.

over Raney nickel, promoted by palladium. The reaction
proceeds with decreasing rate, the initial rate of reaction
being almost independent of the benzaldehyde concentration
in the solution, and fluctuating about 4 ml H_2/min. The
reaction product, benzyl alcohol, inhibits the process, the
progress of which is indicated by the course of the potentiometric curves. The introduction of benzaldehyde into the
reaction mixture contributes to a catalyst potential shift
towards the anodic side by 0.15V, but with a decrease of
benzaldehyde concentration the potential gradually returns
to the cathodic side, although not reaching its initial value
(i.e., the value of the catalyst saturation potential) of
70-80 mV. This is indicative of the reaction product
adsorption, which increases, as well as the catalyst potential shift after the hydrogenation process with the increase
in the benzaldahyde concentration. From the kinetic and
potentiometric curves, the relationships of benzaldehyde
and benzyl alcohol adsorption coefficients (b_1/b_2) and also
the activation energy can be determined. The relationship
b_1/b_2 turned out to be 7, i.e., the reaction product, benzyl
alcohol, adsorbs 7 times weaker than benzaldehyde. The
activation energy, calculated according to formula (1), is
7.5 kcal/mole.
     The hydrogenation reaction rate of benzaldehyde is
proportional to hydrogen pressure up to 20-30 atm. Hence,

in normal conditions over Raney nickel, the hydrogenation
reaction rate is limited by hydrogen activation and inhibited
by the reaction product, i. e., benzyl alcohol.

In order to increase the catalytic activity it is necessary
to strengthen the bond energy between hydrogen and the
catalyst surface; this may be accomplished by introducing
platinum or iridium into the catalyst composition, as well as
by decreasing the reaction product adsorption on the surface
by changing the nature of the solvent. For instance, amine
may be added to the solution. In the case of aniline, which
may form Schiff's base with benzaldehyde, the latter is
drawn off from the catalyst surface into the solution[1] and
the hydrogenation rate increases several times. However,
in such cases the surface "poisoning" with benzyl alcohol
decreases insignificantly. In the case of tertiary amines,
such as dimethyl aniline, the "poisoning" effect decreases
mainly due to benzyl alcohol displacement and alkalization
of the reaction mixture.

A combination of all these factors can produce an important improvement in the hydrogenation process. Numerous
experiments have shown that a conscious change in the
process character and its limiting step could be achieved by
use of following methods:

a) Changing the catalyst chemical composition (promoting,
mixed catalysts, or alloys) (see reports of A. B. Fasman,
K. K. Jardamalyva, A. M. Sokolaya and S. A. Riabinina, and
F. Byjanov).

Figure 2 shows the dependence of the hydrogenation rate
of acetone in water on the composition of a Pt-Ru-catalyst
supported by various carriers. Here we deal with the simplest case, where hydrogenation is accomplished at a steady-
state potential, and the reaction rate is dependent only on the
bond energy of hydrogen with the catalyst surface. An
optimal catalyst always (Fig. 2) contains 40-50 atomic % of
platinum. The hydrogenation rate increases thereby by more
than one and a half an order. The hydrogen-catalyst surface
bond energy (on a Pt-Ru-catalyst) at a saturation potential
differs by 13 kcal/mole (see table); hence the optimal
hydrogen-surface energy bond for acetone hydrogenation
fluctuates between 47 and 48 kcal/mole.

Changes in the structure of the unsaturated compound
involve an essential alteration of the maximum position for
the same catalysts.

As shown in Fig. 3, an optimal catalyst for dimethyl

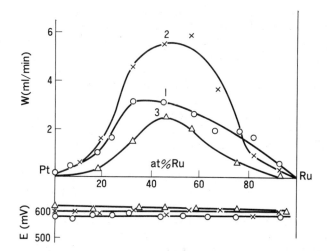

Fig. 2. Dependence of activity and potential of 1 % Ru-Pt catalysts, supported by different carriers, on composition during acetone hydrogenation in water at 20°C; Curve 1, Al_2O_3; 2, ThO_2; 3, TiO_2.

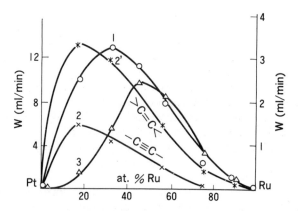

Fig. 3. Dependence of activity of 1 % Ru-Pt/TiO_2 catalysts on composition during hydrogenation of various organic compounds in 0.1 N KOH at 20°C; Curve 1, o-nitrophenol; 2, dimethyl ethynyl carbionol ($C \equiv C$); 2', dimethyl vinyl carbinol (from dimethyl ethynyl carbinol); 3, acetone. (The left-hand scale is for o-nitrophenol, the right-hand one is for DMEC and acetone).

ethynyl carbinol hydrogenation contains 20 atomic % of
platinum, and for o-nitrophenol hydrogenation it contains 30
atomic % of platinum. Unlike acetone both these compounds
are hydrogenated with a considerable potential shift in the
anodic direction in hydrogen-deficient conditions and require
an increased hydrogen-surface bond energy (more than 50
kcal/mole).

The comparison of hydrogenation rates (Fig. 4) with
completely different reaction mechanisms over Pt-Pd
catalysts presents a still more complicate pattern.
Benzoquinone and dinitro-compounds are readily hydrogenated
by catalysts containing 20 atomic % of platinum, the rate-
determining step being the electron transfer from the catalyst
to the unsaturated compound (with anion-radical formation),
n-heptene and dimethyl ethynyl carbinol, the catalyst
containing 5-8 atomic % of platinum. This system confounds
to the general rule that the greater the potential shift, the
more useful is a catalyst with high Me-H bond energy.

Other ways of obtaining optimal catalysts are possible.

a) Supporting the active phase by a carrier which can
readily adsorb one of the components (N. M. Popova and
E. I. Hildebrand).

b) Adsorptive displacement of one of the components,
i. e., selective poisoning (D. V. Sokolsky and A. M. Sokolskaya).

c) Changing the distribution coefficient of the unsaturated
compound between surface and solution by means of solvent
substitution (A. M. Sokolskaya and M. S. Ergeanova).

d) Changing the function of the dissolved hydrogen at
cation adsorption on palladium and nickel (G. D. Zakumbaeva).

e) Regulation of the catalyst potential during hydrogenation (V. P. Schmonina and G. D. Zakumbaeva).

f) Cathodic polarization of the catalyst (nickel) in order
to increase its life (Kirylius and others).

All these methods influence the catalyst activity, selectivity and life. This influence has now been sufficiently
investigated for us to predict in some cases the composition
of optimal catalysts and catalytic systems.[3]

The most recent work in this field has shown that 10-15 %
of alkali earth and rare earth metal oxides present in metal
catalysts the activity of catalysts for hydrogenation in
solutions increased considerably (sometimes by one order).[4]
This is related to the possibility of obtaining optimal catalyst
crystallite sizes and their surface charges.

The critical factor in catalytic hydrogenation is not simply

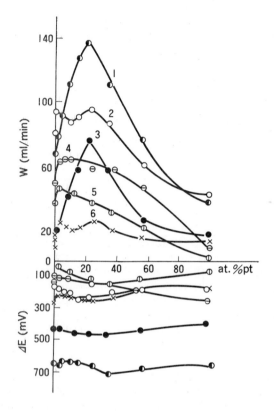

Fig. 4. Effect of composition of Pt-Pd-catalysts, supported by Al₂O₃ (0. 1 g, Me = 4 weight %) on the hydrogenation reaction rate of nitro-compounds (solvent: methanol-dioxane-ammonia), DMEC (water), n-heptene-1 and p-benzoquinone (ethanol) ; curve 1, p-benzoquinone; 2, nitrobenzene; 3, 3, 3'-dinitro-4, 4'-diaminodiphenylethane; 4, n-heptene-1; 5, DMEC; 6, 4, 4'-dinitrodiphenylethane.

the hydrogen-catalyst surface bond energy, but rather the relationship between the bond energies of hydrogen and the carbon atoms of the unsaturated compound. In the case of platinum, this relationship is evidently favourable to hydrogen adsorption (see Table I), and the reaction rate is, in most cases, limited by the activation of the unsaturated compound. The nickel catalyst situation is quite different, the unsaturated compound readily displacing hydrogen from the catalyst surface, thereby forming solid carbide-like species on the surface. That is why the modification of a nickel catalyst is most often concerned with either strengthening the hydrogen-surface bond energy or decreasing the surface adsorption of the unsaturated compound. These modifications are possible by incorporating platinum into the nickel catalyst, supporting the nickel with a carrier readily adsorbing the unsaturated compound and finally, by using a solvent which can form a solvate with the unsaturated compound (A. M. Sokol'skaya and E. N. Bosiakova).

Strong bonds with the carbon atoms of unsaturated compounds are also formed by ruthenium, which becomes a good catalyst with increased hydrogen pressure or when platinum or palladium is incorporated into it.

If we succeed in controlling the degree of the nickel or ruthenium surface coverage with hydrogen, then the catalysts become very good in any conditions of the hydrogenation process in solutions. In the case of palladium it is important to control the work function of the hydrogen atoms (protons) on the surface. This may be accomplished by adsorption of cations on the surface (Ca^{2+}, Zn^{2+}, etc.).

As has already been mentioned, adsorption of the unsaturated compound on the catalyst surface is accompanied by a potential shift towards the anodic side, but an optimal relationship of the reactant species on the surface is often not reached. It should be emphasized, that the catalyst with its adsorbed hydrogen, the hydrogenated compound and its solvent form one common system, the state of which depends on the catalyst nature, the unsaturated compound and the experimental conditions (A. M. Sokolskaya, S. A. Riabinina, D. V. Sokolsky). This system acts in most favorable optimally only under very limited conditions. Therefore, a conscious change of these conditions is required. So, the hydrogenation selectivity of dimethyl ethynyl carbinol in a slightly alkaline solution attains 100 % only at a potential shift of at least 300 mV and a maximum rate is observed even at 200 mV. In the case of mixed catalysts, an optimal potential shift is attainable in the course of the hydrogenation reaction (G. D. Zakumbaeva).

Plotting the charge curves, determination of the electro-conductivity of powder catalysts, plotting a potentio-static curve, performing thermo-desorption or hydrogenation of 3-4 model compounds in a solution, with measurement of the catalyst potential - all the measures cited above permit us to find an answer to the essential question: what does the given hydrogenating catalyst belong to?

In the case of mixed or supported catalysts, all the characteristics cited in Table I vary, and from their appearance (shape) it can be predicted what kind of hydrogen-ation processes the given catalyst is suitable for. As a rule, supported catalysts have a higher hydrogen-surface bond energy.

The degree of adsorption of complex unsaturated organic compounds on palladium is considerably less than on other catalysts. The carrier often changes the adsorption magnitude of the unsaturated compound itself, as well as of the reaction products.

Unfortunately, organic compounds are able to interact with the catalyst atoms in various ways. In a catalytic system simultaneously undergoing hydrogenation reaction, isomerization processes take place, as well as hydrogenolysis, dehydrogenation, polymerization and even destructive processes. In view of this it is necessary to conduct the process in a single direction. Such an approach is character-istic of investigators in the catalysis field, while in electrochemical works, attention is most often focused on ascertaining facts and recording all observed process directions.

In catalysis, to "predict" means to create conditions for the catalytic process to be conducted which will yield a high amount of the required reaction product.

Recently, platinum and rhodium alloys, platinum-ruthenium alloys, promoted Raney nickel catalysts, palladium modified with cations, and other compounds were added to the range of well known catalysts (such as platinum black, calcium oxide supported palladium, Raney nickel, almina-supported nickel, alumina-supported platinum, etc.).

The process of creating new catalysts and catalytic systems is now growing faster owing to our improved understanding of the mechanism of catalytic hydrogenation and the limiting

Table I

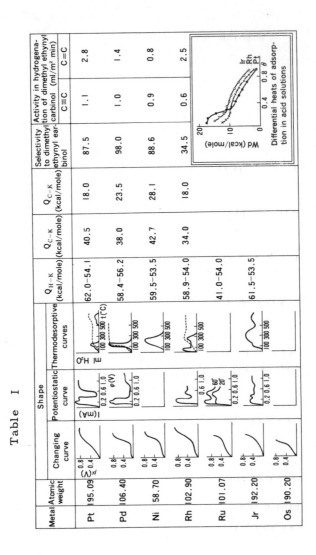

| Metal | Atomic weight | Shape | | | Q_{H-K} (kcal/mole) | Q_{C-K} (kcal/mole) | Q_{C-K} (kcal/mole) | Selectivity to dimethyl ethynyl carbinol | Activity in hydrogenation of dimethyl ethynyl carbinol (ml/m² min) | |
		Changing curve $\eta(V)$	Potentiostatic curve I(mA) $\varphi(V)$	Thermodesorptive curves					$C \equiv C$	$C = C$
Pt	195.09				62.0–54.1	40.5	18.0	87.5	1.1	2.8
Pd	106.40				58.4–56.2	38.0	23.5	98.0	1.0	1.4
Ni	58.70				59.5–53.5	42.7	28.1	88.6	0.9	0.8
Rh	102.90				58.9–54.0	34.0	18.0	34.5	0.6	2.5
Ru	101.07				41.0–54.0					
Jr	192.20				61.5–53.5					
Os	190.20									

Differential heats of adsorption in acid solutions

steps in various conditions, and also owing greatly to the use
of electrochemical methods in the investigation of catalysts
and catalytic processes.

At the moment the following problems remain unsolved:
the influence of hydrogen pressure on reaction rate and
hydrogenation mechanism, and influence of the solvent nature.

References

1) D. V. Sokol'sky, Hydrogenation in solution. Alma-Ata, 1962.
2) D. V. Sokol'sky, Optimal Catalysts for Hydrogenation in Solutions. Alma-Ata, Nauka, 1970.
3) D. V. Sokol'sky, and A. M. Sokol'skaya, Metals as Hydrogenation Catalysts. Alma-Ata, Nauka, 1971.
4) D. V. Sokol'sky, Catalysts for Hydrogenation in Solution. Alma-Ata, Nauka, 1971.

Overvoltage and the Kinetic Law of Elementary Stages of Reaction of Hydrogen Platinum and Nickel Cathode Deposit in Alkaline Solutions

A. Matsuda, R. Notoya and T. Ohmori

Hokkaido University, Sapporo

One of the most important problems in the study of electrochemical processes is the determination of the electrode potential mechanism of origin with relation to an atomic theory.

In the simpliest case of an ideally-polarized electrode potential difference originating on the electrode-solution boundary through a double layer, the electrode potential is determined by the electrode surface free electric charge.

In case of a reversible electrode, this problem has been developed by Professor Frumkin et al.[1-6] at the Institute of Electrohemistry of the USSR Academy of Sciences at Moscow University. They have been studying systems of hydrogen electrodes of a platinum group on the base of Gibbs thermodynamics, and have concluded that the electrode origin is connected not only with the electrode surface free charges, but also with atomic hydrogen and hydrogen ions adsorbing on the electrode surface.

In case of an irreversible electrode the electrode potential change is determined as overvoltage. However, the overvoltage has a difference in free energy between the initial agents and the reaction end-products. Generally, an electrode reaction involves several separate elementary stages. Two questions of interest arise here.

1) In what way is overvoltage distributed between separate stages?

2) Is it possible to explain the overvoltage origination mechanism of origin within the confines of Professor Frumkin's electrode potential theory?

We can answer these questions by investigating an electrode reaction with a galvanostatic method. This paper is concerned with using such a method to study the origin of overvoltage and the kinetic law of separate elementary stages of the reaction of hydrogen platinum and nickel cathode deposit in alkali solutions.[7-11]

273

The galvanostatic method principle consists of giving a direct polarizing current i to an electrode, and measuring the overvoltage change with time. First we shall consider how to analyze the overvoltage curve observed and how to deduce the double layer capacity value and the discharge stage kinetic law.

The moment the current is turned on, overvoltage is caused by electron accumulation on the electrode surface. As soon as the overvoltage begins to increase, the discharge elementary stage of the discharge occurs with a velocity depending on its kinetic law.

If the discharge stage value can be neglected compared with the polarizing current, then the overvoltage increases at a rate which depends on the current density value and the double layer capacity.

If the discharge stage velocity can be approximately compared with a polarizing current, then there is an electron material balance between the polarizing current and the discharge stage and electron accumulation velocities on the electrode surface. In this case we get

$$i_1 = i + C_D \dot{\eta} \tag{1}$$

where i_1 is the discharge stage velocity in electric units, C_D is the double layer capacity and $\dot{\eta}$ the overvoltage change velocity. In the following the overvoltage will be denoted with a minus sign at cathode polarization and the current will be denoted with a plus sign in the cathode direction.

In such case not only the electrode surface free charge but also the intermediate products formed by a discharge stage can participate in the development of the overvoltage.

Even in the case where the double layer capacity value is markedly less than the electrode pseudocapacity value due to intermediate products, the overvoltage may be assumed to be caused by the electrode surface free charge.

As will be shown afterwards in cases of platinum and nickel hydrogen electrodes, the electrode pseudocapacity value is much greater than that of the double layer capacity at cathode polarization.

Under this condition the relation between the overvoltage change velocity and time can be deduced using Eq. (1), provided that the i_1-η relation is known for slight electrode polarization. When η is markedly less than RT/F value, $i_1\eta$ proportionality follows, i.e., we get Eq. (2):

$$\eta = - r_1 i_1 \, ,$$ (2)

where r_1 is a constant which is related to the discharge stage exchange current i_{10} by Eq. (3):

$$r_1 = \frac{RT}{F} \frac{1}{i_{10}},$$ (3)

where R is the gas constant, T the absolute temperature and F, Faraday constant. Equation (4) results from Eqs. (1) and (2):

$$\ln \left(-\frac{i}{\eta} \right) = \frac{t}{\tau_1} + \ln C_D \, ,$$ (4)

where τ_1 is the kinetic constant, a so-called discharge stage time constant, which is expressed by Eq. (5):

$$\tau_1 = C_D r_1 \, .$$ (5)

Equation (4) is valid for an overvoltage curve beginning with any electrode potential.[8] If we observe an overvoltage curve begins with a stationary electrode potential, we can determine C_D and τ_1 values in a stationary state using Eq. (4), but in such a case the value i in Eq. (4) should be replaced by the current density increment Δi. Then r_1 is expressed as a differential coefficient by Eq. (6):

$$r_1 = -\frac{d\eta}{di_1} = f(i_s, \ \eta_s) \, ,$$ (6)

and is a function of the current density i_s or overvoltage η_s in a stationary state.

On the base of Eq. (6) we get the overvoltage value η_{is} , which has resulted from the electrode surface free charge at electrode stationary polarization. Integrating Eq. (6) we get Eq. (7):

$$\eta_{is} = - \int_0^{is} r \, di_s \, ,$$ (7)

since the value i_1 is equal to the value i_s at electrode stationary polarization. With Eq. (7) we can calculate from the overvoltage curves the discharge stage kinetic law in stationary states separately from other elementary stages.

Later we shall present our experimental results from a galvanostatic method in systems of platinum and nickel hydrogen electrodes in alkali solutions.

Figure 1 shows log (- Δi/$\dot{\eta}$)-time ratio on an evaporated platinum film in a cesium hydroxide solution, pH 10.86~9. These ratios are derived from the overvoltage curves beginning with various stationary states in which the over- voltage value is in the range of 0-200 m.

From the figure it is seen that log (-Δi/$\dot{\eta}$)-time ratio is expressed by Eq. (4). The curve slope coefficient in this figure has the value τ_1, which decreases as the overvoltage is increased.

The next point to note is that the extrapolations of these curves meet at a single point on the ordinate. This point has a double layer capacity value. Therefore, it follows that this value remains constant and does not depend on the overvoltage value.

It was also found that the value of C_D remains constant while changing the solution composition and concentration.

However, even the initial part of the curve deviates from a rectilinear relation. This deviation may result from the electrode surface irregularities, since it depends on the

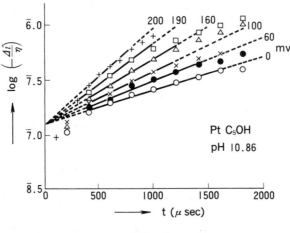

Fig. 1

electrode surface state.

A similar relation was also found for nickel in alkali solutions.

Figure 2 shows the log $(-\Delta i/\dot{\eta})$-time ratio on a nickel-evaporated film in 0.01 N sodium hydroxide solution.[10]

We have seen that this ratio is expressed by Eq. (4), and that the value τ_1 decreases with increasing the overvoltage, and that the value C_D remains constant and does not depend on the overvoltage value.

Later we shall consider the kinetic parameter τ_1 change with varying ionic concentrations of the alkali metal C_{M^+}.

Figure 3 shows log τ_1-values as a function of log C_{M^+} in sodium and cesium ion solutions.[8]

The white and black symbols denote the hydroxide and sulphate solutions, respectively. The pH values of the sulphate solutions are in the range of 8.2-11.4. It is seen from the figure that log τ_1 decreases with increasing log C_{M^+} rectilinearly with a gradient coefficient equal to -1/2 and does not depend on the solution pH.

Figure 4 shows a similar log τ_1 - log C_{Nd} relation for nickel at equilibrium in hydroxide and sodium sulphate solutions.[11] The pH-values of sodium sulphate solutions are in the range of 8.1-9.9. It can be seen from the figure that log τ_1 also decreases with increasing log C_{M^+}, with the gradient -1/2, and this increase is independent of the solution pH.

From this it follows according to Eqs. (3) and (5) that the discharge stage exchange current varies proportionally to the square root of the alkali metal ion concentration and does not depend on the solution pH. Since the value of C_D remains constant while the solution concentration is being varied, we get Eq. (8):

$$i_{10} = k \ (C_{M^+})^{\frac{1}{2}} \ , \tag{8}$$

where k is the discharge stage velocity constant.

From this we concluded that an elementary stage is formed from the alkali metal discharge in systems of platinum and nickel hydrogen electrodes in alkali solutions.

Afterwards we shall consider the kinetic parameter γ_1 at electrode stationary polarization which is determined from the values C_D and τ_1 with Eq. (5).

The values γ_1 for platinum are given in Fig. 5 as a function of a polarizing current i_s in cesium hydroxide solutions

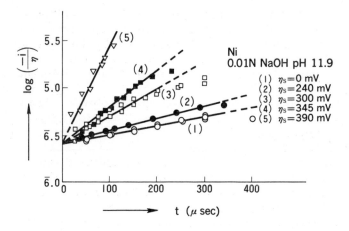

Fig. 2

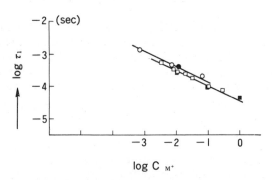

Fig. 3

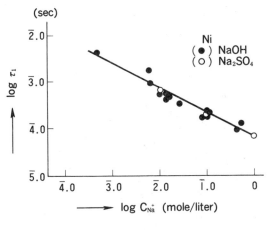

Fig. 4

of different concentration.[9)]

In a diluted solution of pH 10. 86, the value r_1 markedly decreases with increasing values of i_s while in a solution of pH 12. 80, the value r_1 remains approximately constant with varying i_s.

Integrating r_1 -vs. - i_s curves in Fig. 5 on the base of Eq. (7) we arrive of the overvoltage value η_{is}, caused by a free charge at i_s — the polarizing current density.

In Fig. 6, the ratio of log i_s to η_{is} is given by curve 1 at electrode stationary polarization in a solution of pH 10. 86.

Note that this ratio is expressed by Frumkin kinetic equation with a Tafel constant of 0.5, in which a potential member ψ is neglected:

$$i_s = i_{10} \exp\left(-\frac{F\eta_{is}}{2RT} \right) - \exp\left(\frac{F\eta_{is}}{2RT} \right) \qquad (9)$$

For comparison, the polarization curve of the overall reaction, log i_s against η_s, in the same solution is also given in Fig. 6 (curve 4).

It can be seen by comparing curve 1 with curve 4 in Fig. 6 that at low current densities there is an overvoltage region where the value η_{is} is much less than the value η_s, and in fact is equal to zero.

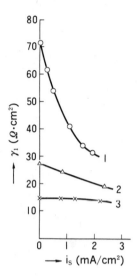

Fig. 5

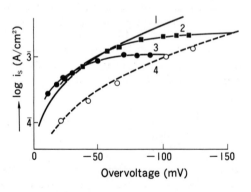

Fig. 6

The η_s - η_{is} difference at a certain constant value i_s, which we denote by η_{2s}, shows the value of the overvoltage caused by intermediate products at electrode stationary polarization.

Next we shall consider the overvoltage in the course of transition from an equilibrium potential to a stationary state in accordance with a constant polarizing current, the discharge stage velocity in any electrode potential being expressed on the basis of the electrode material balance by Eq. (1). The relation between log i_1 and η- determined in such a way in the course of the overvoltage transition is shown by curves 2 and 3 in Fig. 6.

Curves 2 and 3 both consist of two regions, the first of which coincides with Curve 1, and in the second region the overvoltage continues to increase even though the discharge stage velocity has almost reached a polarizing current value.

From these facts it follows that while increasing the overvoltage after the current has been turned on, the overvoltage caused by a free charge arises first and then the overvoltage caused by intermediate products arises, i. e., these two types of overvoltage arise separately during the overvoltage transition.

A similar relation was also found for other concentrations of a cesium hydroxide solution. Fig. 7 shows the curves of the discharge stage polarization and of the overall dehydration reaction at electrode stationary polarization and during the overvoltage transition in a solution of pH 11. 86.[9] At low current densities the value η_{is} is practically equal to zero.

Figure 8 shows a similar relation with platinum in a cesium hydroxide solution of pH 12. 80.[9]

The discharge stage polarization curve is separated from the other curves, and at low current densities the value η_{is} is practically equal to zero.

In Fig. 9 are shown the above-mentioned discharge stage polarization curves in cesium hydroxide solution of various concentations at electrode stationary polarization. The discharge velocity at a certain constant value, η_{is}, increases with increasing the solution concentration, since the discharge stage is formed from the alkali metal ion discharge.

Broken line (curve 1) in Fig. 9 shows the relation between log i_s and η_{2s} at electrode stationary polarization[9] , where η_{2s} denotes η_s - η_{is} . It is seen that this curve does not depend on the solution pH, and therefore demonstrates

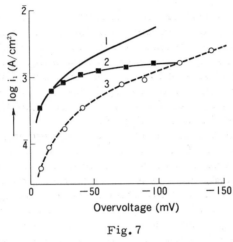

Fig. 7

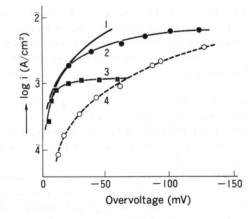

Fig. 8

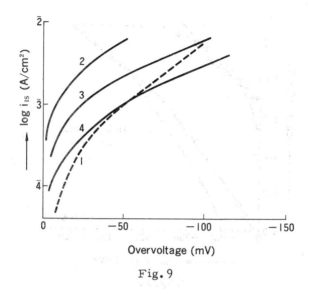

Fig. 9

the kinetic law of the recombination elementary stage of hydrogen atoms being adsorbed on to the electrode surface.

A similar relation was also found for nickel in sodium hydroxide solutions. Curve 1 in Fig. 10 shows the discharge stage polarization curve of log i_s against η_{1s} and curve 2 shows the overall reaction polarization curve at stationary polarization in a sodium hydroxide solution of pH 12.80.[10] Curve 1 follows the Frumkin equation. A black circle shows the log i_1 -vs.- η ratio during the overvoltage transition after the current has been turned on. It is seen that log i_1, η - curve consists of two regions as in the case of platinum. A white triangle shows the log i_1 -vs.- η ratio during the overvoltage decrease after the current has been turned off. In this case i_1 is determined by the $i_1 = C_D\eta$ equation. At the initial part of the overvoltage decrease, log i_1 determined in such a way decreases with decreasing the overvoltage within the range of the log i_s - η_s curve. From this it follows that in this overvoltage range at stationary polarization the overvoltage increase is due to the electron accumulation on the electrode surface and this overvoltage region falls initially after the polarizing current has been

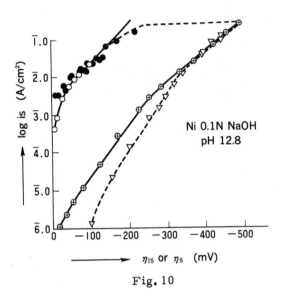

Fig. 10

turned off.

In Fig. 11, log i_s is expressed as a function of η_{1s} and η_{2s}.[10] Curves 1, 2 and 3 show the relation between i_s and η_{1s} and curves 1', 2' and 3' show same relation, but in sodium hydroxide solutions of different concentrations. It is seen from the figure that the log i_s -vs.- η_{2s} curves are almost independent of the solution pH. Therefore, these curves also demonstrate the kinetic law of the recombination stage of hydrogen atoms, as in case of platinum. At low current densities the value η_{1s} is practically equal to zero independently of η_s, as in case of a platinum electrode.

From these experimental results it can be concluded that the hydrogen cathode reaction on platinum and nickel in alkaline solutions follows the scheme:

$$M^+ + e \longrightarrow M \tag{I}$$
$$M + H_2O \longrightarrow H + OH^- + M^+ \tag{II}$$
$$H + H \longrightarrow H_2 \tag{III}$$

where M and H denote, respectively, the alkali metals and hydrogen atoms adsorbed on the electrode surface.

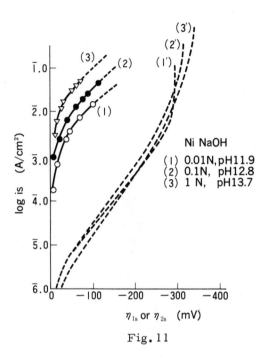

Fig. 11

At low current densities the discharge stage is practically in equilibrium, as η_{1s} is equal to zero. The water molecule decomposition stage is always in quasiequilibrium as the overall reaction overvoltage η_s involves the discharge overvoltage and the atomic hydrogen recombination overvoltage.

When η_{1s} is equal to zero the free charge density ξ of the electrode surface remains constant, i.e.,

$$\Delta \xi = O \qquad ; \qquad (10)$$

as η_{1s} is caused by varying electrode surface free charge.

Next we shall consider which of the intermediate products, M or H, plays the main part in the overvoltage establishment η_{2s}. For this purpose we observed the electrode pseudocapacity from the curves of the overvoltage

fall after the current had been turned off. Figure 12 shows
the platinum electrode pseudocapacity values in 1 N sodium
sulphate solutions of pH 4.7 and 11.4, and in a 0.01 N
solution of pH 8.2 as a function of the electrode potential φ
in relation to the normal hydrogen electrode potential.[7]

From this figure it is seen that the electrode pseudocapacity has a maximum value at approximate potential of
about - 800 MB, and that the value is independent of the
solution pH, but increases with increasing alkali metal ion
concentration.

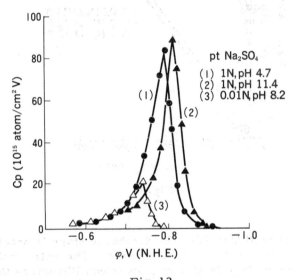

Fig. 12

Figure 13 shows a similar relation for nickel in sodium
sulphate solutions of various pH values from 9.80 to 12.80.[10]
The electrode pseudocapacity also has a maximum value at
electrode potential near to - 900 MB, and this value is
independent of the solution pH.

Therefore, it follows that the intermediate atom of the
electrode surface alkali metal changes its concentration
with varying overvoltage, and the electrode surface atomic
hydrogen does not cause the electrode pseudocapacity
origination at cathode polarization. In other words the
intermediate alkali metal plays the main part in the over-

voltage establishment in the case of platinum and nickel
hydrogen electrodes in alkali solutions.

Later we shall consider the overvoltage origination
mechanism in the systems under investigation on the basis
of the Frumkin electrode potential theory.[1.8)]

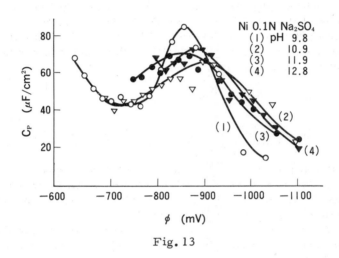

Fig. 13

As it is seen from the comparison of the polarization
curves of the cathode hydrogen deposite elementary stages a
slow stage at low current densities is the atomic hydrogen
recombination and the discharge stage and the water molecular decomposition stage are in quasiequilibrium, i.e., we
get

$$M^+ + e \rightleftharpoons M \qquad\qquad (I')$$
$$M + H_2O \rightleftharpoons H + OH^- + M^+ \qquad\qquad (II')$$

Under these conditions the overvoltage origination mechanism can be determined within the limits of the Frumkin
theory on the basis of Gibbs. thermodynamics. From the
discharge stage thermodynamic equilibrium we get

$$d\varphi = d\mu_{M^+} - d\mu_M \qquad (11)^*$$

In case of the systems under consideration the following conditions are typical.

1)   Cathode polarization atomic hydrogen does not take part in the overvoltage establishment, but the electrode surface alkali metal is the origin of the overvoltage. Under these conditions the electrode does not act as a hydrogen electrode, but as an alkali metal electrode, i. e., the electrode potential φ is determined by the alkali metal atom concentration A_M and by the density of the electrode surface free charge, i. e., we get

$$d\varphi = X dA_M + Y d\xi$$

$$X \equiv \left(\frac{\partial \varphi}{\partial A_M}\right)_\varepsilon = -\frac{1}{C_D}, \quad Y \equiv \left(\frac{\partial \varphi}{\partial \xi}\right)_{A_M} = \frac{1}{C_D} \qquad (12)$$

where the differential coefficients X and Y are equal, respectively, to the inverse values of the electrode pseudo-capacity C_P and the double layer capacity C_D, which can be determined by a galvanostatic method.

2)   At low current densities the density of the electrode surface free charge remains constant and does not depend on the electrode potential:

$$\xi = \text{const.}$$

In 1936, A. Shlygin, A. Frumkin, and W. Medvesovsky and more recently, O. A. Petrii et al.[4,5]   showed that the free charge density of the hydrogen platinum electrode surface remains constant during anode polarization in alkali solutions. This was confirmed by N. A. Balashova and V. E. Kazariov[12] by measuring ion adsorption on the electrode surface with a radiochemical method. Our results show that the charge density of the platinum and nickel electrode surface also remains constant during cathode polarization by low current densities.

3)   However, anion adsorption on platinum can be neglected at anode polarization potentials near to the elec-

* Chemical potentials μ_{M^+} and μ_M are expressed in electrical units.

trode equilibrium potential.[12] From this it can be assumed that anion adsorption does not take place on platinum during cathode polarization, because the electrode surface free charge is becoming still more negative. We get

$$\sqrt{A^-} = O.$$

Under conditions 1), 2) and 3), the mathematical consideration based on the Frumkin theory leads to the following results:

$$XdA_M = -d\mu_M + \frac{X}{X + Y} d\mu_{M^+} \tag{13}$$

$$Yd\xi = \frac{Y}{X + Y} . \tag{14}$$

In the case of the systems under consideration, $|X| \ll Y$ or $C_p \gg C_D$, and so we get

$$XdA_M = d\mu_M = d\varphi_{20} \tag{15}$$
$$Yd\xi = d\mu_{M^+} = d\varphi_{10} . \tag{16}$$

These relationships show that the electrode potential in the systems under consideration is divided into two independent components, one of which is caused by the electrode surface alkali metal which depends only on μ_M and the other caused by the electrode surface free charge which depends only on μ_{M^+}. In the second case the electrode surface charge density does not depend on the solution pH.

Let us denote two members of the electrode potential components in the systems under consideration as φ_{10} and φ_{20}, respectively. Then from Eqs. (15) and (16) we get

$$\varphi_{10} = \frac{RT}{F} \ln a_{M^+} + const. \tag{17}$$

$$\varphi_{20} = -\frac{RT}{F} \ln a_M + const. \tag{18}$$

The form of Eq. (17) is consistent with the Nernst equation.

If $d\varphi_{20}$ is caused by a dipole moment μ between atoms adsorbed on the surface of the electrode and alkali metal, then $d\varphi_{20}$ can be expressed as

$$- d\varphi_{20} = 4\pi A_M \mu d\theta \tag{19}$$

where θ denotes the alkali metal atom filling. From Eqs. (18) and (19) it follows that the alkali metal adsorption isotherm is expressed as a semi-logarithmic isotherm, a so-called Frumkin-Temkin isotherm:

$$\theta = \frac{1}{f} \ln a_M \tag{20}$$

But in the case of platinum we observed that the sodium filling of the electrode surface greatly exceeds unity. From this it follows that a real intermetallic compound is formed near to the electrode surface, as was shown by Professor Kabanov in the case of some other metals in alkali solutions. In this case the potential results from the difference in the change of the contact potential between the layers of surface compound and basic electrode metal.

Next we shall consider the discharge stage kinetic law taking into consideration the overvoltage separate components.[8]   The discharge velocity is expressed by the Frumkin equation as was shown in the experimental section:

$$i_1 = k_I + a_M + \exp\left(-\frac{\alpha F \varphi_1}{RT}\right) - k_{I^-}\exp\left(\frac{\beta F \varphi_1}{RT}\right) \tag{21}$$

where φ_1 denotes the electrode potential component caused by the electrode surface free charge.

Equation (21) is rewritten in the form:

$$i_1 = i_{10}\left\{\exp\left(-\frac{\alpha F \eta_1}{RT}\right) - \exp\left(\frac{\beta F \eta_1}{RT}\right)\right\} \tag{22}$$

where

$$i_{10} = k_I + a_M + \exp\left(-\frac{\alpha F \varphi_{10}}{RT}\right) = k_{I^-}\exp\left(\frac{\beta F \varphi_{10}}{RT}\right) \tag{23}$$

and

$$\eta_1 = \varphi_1 - \varphi_{10} \tag{24}$$

If Eq. (17) is substituted for Eq. (23), the discharge stage exchange current takes the form:

$$i_{10} \propto (d_{M^+})^{1-\alpha} \propto (d_{M^+})^{\frac{1}{2}} \tag{25}$$

Since $\alpha = \beta = 1/2$, i_{10} is proportional to the square root of the alkali metal ion concentration in accordance with the experimental result.

Equation (21) can be rewritten in the following form using the electrode potential φ:

$$i_1 = k_{I^+} a_{M^+} \exp\left(-\frac{\alpha F(\varphi-\varphi_2)}{RT}\right) - k_{I^-} \exp\left(\frac{\beta F(\varphi-\varphi_2)}{RT}\right) \tag{26}$$

where φ_2 denotes the component of the electrode potential, which is caused by the alkali metal intermediate atom. If Eq. (18) is substituted for Eq. (26), we get

$$i_1 = k_{I^+}'' a_{M^+} a_M^{-\alpha} \exp\left(-\frac{\alpha F\varphi}{RT}\right) - k_{I^-}'' a_M^{\beta} \exp\left(\frac{\beta F\varphi}{RT}\right) \tag{27}$$

From Eq. (27) it follows that in equilibrium we get

$$\varphi = \frac{RT}{F} \ln a_{M^+} - \frac{RT}{F} \ln a_{M^-} \text{ const.} \tag{28}$$

in accordance with a thermodynamic relation.

If the alkali metal activity a_M is expressed by $e^{f\theta}$, in accordance with a semi-logarithmic isotherm, then from Eq. (27) we get

$$i = k_{I^+}'' a_{M^+} \exp(-\alpha f\theta) \exp\left(-\frac{\alpha F\varphi}{RT}\right) - k_{I^-}'' \exp(\beta f\theta) \exp\left(\frac{\beta F\varphi}{RT}\right) \tag{29}$$

Equation (29) is in agreement with the equation of a proton discharge stage which was deduced by Frumkin and Aladjalova in 1944,[13] provided that the activities M^+ and M are replaced by the proton and atomic hydrogen activities, respectively.

In conclusion, we repeat that in the systems of platinum and nickel hydrogen electrodes in alkali solutions the hydrogen cathode deposit overvoltage is divided into two independent components with the help of a galvanostatic method, and these two independent components are

expressed on the basis of the Frumkin electrode potential
theory as a function of the ion and atom chemical potential
of an alkali metal. The discharge stage velocity depends only
on the electrode potential component caused by the electrode
surface free charge and its law follows the Frumkin
equation.

SUMMARY

This paper is concerned with a description of our studies
of the overvoltage origin and the kinetic law of separate
elementary stages of the reaction of hydrogen, platinum and
nickel cathode deposit in alkaline solutions by a galvanostatic
method.

The following conclusions have been reached on the basis
of experimental results. The overvoltage η of an overall
reaction in the systems investigated is divided into two
independent components η_1 and η_2 by a galvanostatic
method. One of the components, η_1 , is expressed by the
electrode potential difference $\varphi_1 - \varphi_{10}$ caused by the electrode surface free charges between stationary and equlibrium
states, and the other component, η_2 , is expressed by the
electrode potential difference $\varphi_2 - \varphi_{20}$ caused by the intermediate atoms of the electrode surface alkali metal between
stationary and equilibrium states.

The reaction of hydrogen cathode deposit in these
systems follows the scheme: $M^+ + e \longrightarrow M$; $M + H_2O \longrightarrow$
$H + OH^- + M^+$; $H + H \longrightarrow H_2$, where M^+ is an alkali metal
ion, and M and H are alkali metal and hydrogen atoms
adsorbed on the electrode surface. At slight electrode
slight polarization atomic hydrogen recombination is a slow
stage, other stages being in equilibrium. Next it is shown
that η_1 become a discharge stage overvoltage and η_2 is
responsible for the overvoltage of the atomic hydrogen
recombination stage. The discharge stage velocity is
expressed by the Frumkin equation as a function of the
potential φ_1.

At slight electrode polarization the mathematical consideration of the electrode potential origin on the base of the
Frumkin electrode potential theory leads to the following
results:

$$\varphi_{10} = (RT/F)\ln a_{M^{++}} + \text{const.} \tag{1}$$
$$\varphi_2 = -(RT/F)\ln a_M + \text{const.} \tag{2}$$

where a_{M^+} and a_M denote the alkali metal ion and atom activities, respectively. The form of Eq. (1) is consistent with the Nernst equation.

Equation (2) leads to a semi-logarithmic isotherm, a so-called Frumkin-Temkin isotherm, provided that the electrode potential is due to a dipole moment between the atoms of the electrode surface alkali and electrode base metals. It was observed that the atom filling of the electrode surface alkali metal greatly exceeds unity. In this case, Eq. (2) shows the contact potential difference between the layers of the surface compound and electrode base metal and the relation between the electrode potential and the activity of the electrode surface alkali metal atom.

References

1) A. N. Frumkin, Elektrokhimija, 2, 387 (1966).
2) A. Slygin and A. N. Frumkin, Acta Physicochim., U. R. S. S., 3, 791 (1935).
3) A. Slygin, A. N. Frumkin, and W. Medwedowsky, ibid., 4, 911 (1936).
4) O. A. Petrii, A. N. Frumkin, and Yu. G. Kotlov, Elektrokhimija, 5, 476, 735 (1969).
5) O. A. Petrii and Yu. G. Kotlov, Elektrokhimija, 4, 774 (1968).
6) A. N. Frumkin, N. A. Balashova, and V. E. Kazarinov, J. Electrochem. Soc., 113, 1011 (1966).
7) A. Matsuda and R. Notoya, J. Res. Inst. Catalysis, 14, 165, 192, 198 (1966).
8) A. Matsuda and R. Notoya, ibid., 18, 53, 59 (1970).
9) R. Notoya, ibid., in printed.
10) T. Ohmori and A. Matsuda, ibid., to be published.
11) T. Ohmori and A. Matsuda, ibid., 15, 247 (1968).
12) N. A. Balashova and V. E. Kazarinov, Uspekhi Kjimii, 34, 1721 (1965).
13) A. Frumkin and N. Aladjalova, Acta Physicochim., U. R. S. S. 19, 1 (1944).

On the Mechanisms of Enzymatic Reactions

O. M. POLTORAK and E. S. CHUKHARAI

Moscow State University, Moscow

The mechanisms of many enzymatic reactions are studied now in terms of catalytic mechanisms, which they closely resemble. Owing to the success of X-ray diffraction methods, the elementary acts are now sometimes studied better in enzymatic reactions than in catalytic reactions. This fact makes it possible to analyze the general properties of the enzymatic reactions.

One of us[1] has presented evidence that dissimilar catalytic and enzyme reactions may have common general properties. This evidence was based on the concept of the bond redistribution chains in the enzymatic globulas (BRC). Here we shall present some supplementary data to support this idea.

In many biochemical studies it has been shown that the catalytic process in the protein globula is connected with appreciable changes in its geometrical, physical and chemical properties.[2] As a rule, the dozens of atoms and atomic groups are involved in catalysis. They establish the necessary close environment for the catalytic groups and orient the reactants. However, it has been shown that this rather complicated picture can be greatly simplified[1] if we begin to analyze the enzymatic reactions considering only the chemical changes during the act of catalysis, and setting aside all the changes which take place during the adsorption of a substrate by the protein globula.

To underline the significance of the chemically unchanged atomic environment of the enzyme globula, we are discussing here only the specific question of chemical mechanisms of enzymatic reactions. Only in this sense are the enzymatic reaction mechanisms considered below.

Our most interesting result concerned with the "chemical mechanism" is the clarification of many catalytical mechanisms for quite different enzymatic reactions in which different reactants, catalysts and prosthetic groups are used. The majority of the general properties of the enzymatic reactions involve acid-base mechanisms. Addition or elimination of protons in the linear bond sequence involved in act

of catalysis is ruled not only by enzymatic hydrolysis but also
inevitably by the oxidation-reduction processes such as the
hydro-dehydrogenation, the oxidation of amines or amino
acids by molecular oxygen, the decomposition of H_2O_2, etc.
As for the others, effective elementary acts of the linear
chain of the bond redistribution (BRC) is formed. Now we
begin to consider the properties of the BRC for enzymatic
reactions of different classes.

Hydrolases and Transferases Without Prosthetic Groups

   To this type of enzymes belong numerous groups of
protein molecules, of which only the amino acid residues off
polypeptide chains are catalysts. These active residues include
the imidazol ring of histidine, the carboxyl groups of aspara-
ginic or glutamic acid, the phenol group of tyrosin, the -SH
group of cystein, the -OH group of serine and the -NH_2 group
of lysine. The active centers of hydrolases and transferases
always contain two or three such groups. This is important,
since it confirms the proposed mechanisms for the chemical
changes in BRC of the enzymes in question. Three enzymes
studied by X-ray diffraction (with a resolution of 2 Å) are the
hydrolases lysozyme,[3] α-chymotrypsin[4] and related proteins
and carboxypeptidase.[5]

   To illustrate the mechanism of catalysis by BRC in the
globula, we shall consider first the properties of enzymes of
known structure. The chemical mechanism of the hydrolysis
of the peptide bond by carboxypeptidase may, according to
Lipscomb,[5] be represented as Fig. 1.

   The adsorption centre of the enzyme which contains Zn-ion,
imidazol ring and a hole for the peptide here is not shown
here, since they are not changed during catalysis. The
catalytic reaction involves the redistribution of the six bonds
originally forming a linear sequence. If the absence of
covalent bonding is denoted by zero multiplicity (0), the
main property of the BRC may be described according to a
very simple rule: catalytic reaction is going on the alternative
change in the multiplicities of chemical bonds forming the
BRC, by +1 and -1 (or -1 and +1), along the sequence of bonds
involved in act of catalysis. This will be called the "rule 1".
If we take into account the atoms in the BRC starting from the
oxygen atom of the Glu 270, the chemical mechanism of this
reaction may be represented by the following equation:

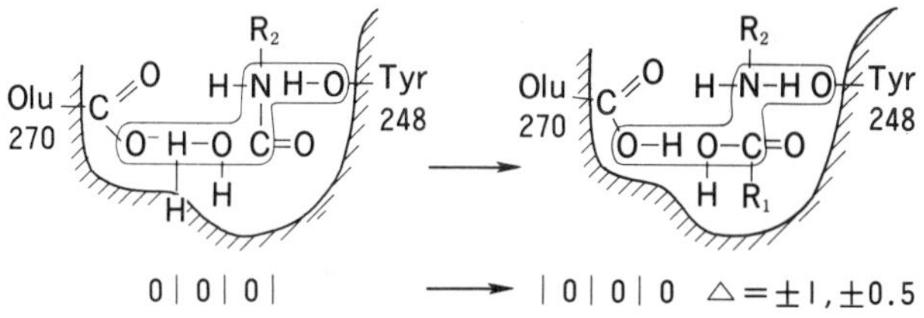

Fig. 1.   Mechanism of oxygen incorporation into acetone.   The charge of each
species is omitted in the figure.
H_a, OH_a and O_a do not necessarily mean
adsorbed hydrogen, hydroxyl, and oxygen
radicals. O_a represents the active species
drived from molecular oxygen.   It also
includes the active species from the bulk
oxide.

$$(\text{Glu270})O^-\;H\text{-}O\;C\text{-}N\;H\text{-}O(\text{Tyr248}) \rightarrow (\text{Glu})O\text{-}H\;O\text{-}C\;N\text{-}H\;O^-(\text{Tyr})$$

$$0 \quad 1 \quad 0 \quad 1 \quad 0 \quad 1 \qquad\qquad 1 \quad 0 \quad 1 \quad 01 \quad 0 \varDelta = \pm 1$$

$$\tag{1}$$

The upper line shows the atoms involved in the chemical
changes and the lower line shows the multiplicities of the
corresponding bonds.   Similar linear bond sequences with
analogous properties are found in other enzymatic reactions, and they always possess the properties of the linear
BRC.   By definition, and BRC exists in the two forms
"initial" and "final" the transitions between which are
described by the ± 1 rule.
     The BRC transformation due to carboxypeptidase is
connected with two proton transitions to the neighboring
electronegative atoms, when the two covalent hydrogen
bonds change places, with the subsequent redistribution
of electron densities in other parts of the BRC.   The total
negative charge is transported from one catalytic group to
the another by the reaction (from Glu 270 to Tyr 248), but
this is not a simple transfer of electron through the BRC.
The system of the bonds in BRC is a cooperative system,
but it is possible to recognize three steps in the chemical
change in the BRC.   Two of these are connected with the

proton transitions. For the same BRC it is possible to represent the scheme as follows:

substrate
↓

$$,\text{O}^- \quad \text{H-O} \quad \text{C-N} \quad \text{H-O} \longrightarrow \text{O-H} \quad \text{O}^- \quad \text{C-N} \quad \text{H-O} \qquad (2)$$

(Glu 270) ↑ (Tyr 248)
(H$_2$O)

$$\text{O-H} \quad \text{O}^- \quad \text{C-N} \quad \text{H-O} \longrightarrow \text{O-H} \quad \text{O}^- \quad \text{C-N-H} \quad \text{O}^- \qquad (3)$$

$$\text{O-H} \quad \text{O}^- \quad \text{C-N-H} \quad \text{O}^- \longrightarrow \text{O-H} \quad \text{O-C} \quad \text{N-H} \quad \text{O}^- \qquad (4)$$

The cooperative properties of the BRC are connected with the fact that in the enzyme globula any of the steps are more effective than the same steps in isolated homogenous reactions.

We are able to show some properties of the BRC common to many reactions.

In the BRC there is only slight displacement of nuclei to neighboring positions. Redistribution of the protons in the BRC is not connected with the occurrence of free radicals. Atoms of O and N, among which the proton transitions occur, possess the necessary properties for this. They increase their valency by 1 ($^{III}\text{O}^+$ and $^{IV}\text{N}^+$) by accepting protons. After the proton elimination they decrease the valency again by 1 ($^{I}\text{O}^-$ and $^{II}\text{N}^-$).

In the literature there are descriptions of the mechanism for the action of the α-chymotrypsin. The structural data on this enzyme confirms the mechanism proposed by Blow, Birktoft and Hartly,[4] but the properties of the hydroxyl group of serine (ionized serine group at pH~8) are unusual from the point of view of pure chemistry. However, the mechanism proposed by Bender is not in agreement with the structural data suggesting a different orientation of the imidazole ring to the molecule of substrate). We shall discuss both of these possible mechanisms.

According to Blow,[4] the chemical mechanism of the hydrolysis of peptide bond by α-chymotrypsin during the formation of the acyl serine in enzyme globula, may be represented as Fig. 2. In this case it is easy to see the BRC, in which the O atom of carboxyl group of Asp 102, a part of the imidazole ring of His 57, and the C-N bond of the substrate and the oxygroup of serine 195 takes part. The BRC has 8 rings, the bonds of which are continuously being

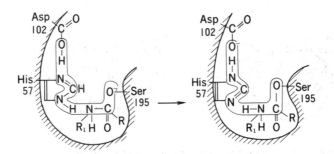

Fig. 2. Possible mechanisms for acetone formation.

broken and reformed:

(Asp 102) O-H N=C-N-H N-C O⁻ (Ser 195) ⟶
(Asp) O⁻ H-N=C H-N C-O (Ser 195)

The change in the multiplicities of the chemical bonds in the act of formation of acyl-α-chymotrypsin may be written in this form, starting from Asp 102:

$$1\ 0\ 2\ 1\ 1\ 0\ 1\ 0 \longrightarrow 0\ 1\ 1\ 2\ 0\ 1\ 0\ 1\ \mathit{\Delta} = \pm 1$$

with the alternating increasing and decreasing by unity of the multiplicities of the bonds in the sequence. In the carboxy-peptidase reaction the change in the BRC is due to the two proton transitions.

The chemical mechanism of the same process according to Bender may be represented as Fig. 3. In this case BRC takes the form of a cycle with the same general properties as before. The transition to acyl-serine complies with the rule $\mathit{\Delta} = \pm 1$. (Fig. 4.). The chemical mechanism of the reaction accelerated by ribonuclease (the formation of cyclophosphate) may be written in the form shown in Fig. 5.

The high degree electrical and chemical symmetry of BRC for this reaction is remarkable.

$$N^{+}\text{-H O-P O-H N} \longrightarrow N\ H\text{-O P-O H-}N^{+}$$
$$1\ 0\ 1\ 0\ 1\ 0 \qquad\qquad 0\ 1\ 0\ 1\ 0\ 1 \qquad \mathit{\Delta} = \pm 1$$

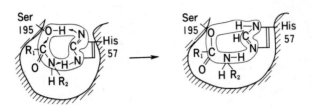

Fig. 3.   Mechanisms of acrolein formation on MoO_3-Bi_2O_3 and Pd-activated
charcoal catalysts.

Fig. 4.   Oxidation of propylene over
SnO_2-MoO_3 (A) catalyst. Catalyst used:
17.2 g. GHSV: 660 ml-STP/ml-cat.hr.
Amounts of products are normalized by
carbon balance.
    A small amount of isopropyl alcohol
(2-3μ mole /min) is formed at each
temperature.

The BRC may be represented by only one symbol, which we
have to read from left to right for the initial situation and
then finally from right to left. The BRC in the enzyme-
substrate complex of lysozyme is also symmetrical. For the
basic general mechanism of hydrolysis, the chemical mechanism in this case be represented in the form shown in
Fig. 6.
    The BRC is as follows:

(Glu 35) O-H O-C O-H O⁻ (Asp 52) ⟶
        1   0   1   0   1   0

(GLU) O⁻ H-O C-O H-O (Asp)
       0   1   0   1   0   1 $\Delta = \pm 1$,

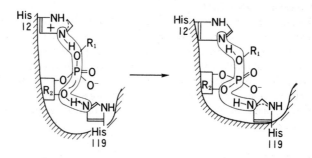

Fig. 5

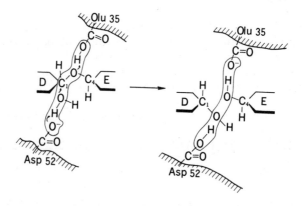

Fig. 6

The simplest, though symmetrical, BRC corresponds to the first mechanism of this reaction proposed by Phillips:

$$(\text{Glu } 35) \quad \text{O-H} \quad \text{O} \quad \longrightarrow (\text{Glu}) \quad \text{O}^- \quad \text{H-O}^+$$
$$1 \quad 0 \qquad\qquad\qquad 0 \quad 1 \qquad \Delta = \pm 1$$

However, the BRC is not symmetrical in all cases. It is

altered by the difference in bond lengths for O-H and O H, and is also dependent on the various environments of the different parts of the BRC.

We have now related all we know about the mechanisms of this kind of enzymatic reaction on the basis of the precise structural data (2Å resolution). The results of the chemical and kinetic investigations permits us to regard our proposed mechanisms for the rather different enzymatic reactions as being more or less reliable. Some of these are detailed in refs. 1 and 2, and in all these cases the same properties of the BRC were found.

Dehydrogenases without a Prosthetic Group

The large groups of the oxidoreductases are formed by the enzymes of **hydro-dehydrogenation processes** using the specialised substrates of the reactions——NAD^+-NADH and $NADP^+$-NADPH. The total reactions of hydrogen transport have the form:

$$NADH + A + H^+ \longrightarrow NAD^+ + AH_2$$

and occur in the presence of corresponding enzymes. The catalytic groups of these enzymes are the acid-base groups of the protein molecule. The mechanism of the catalysis includes the reversible transport of a proton from catalytic group and an hydride ion (H^-) from NADH to the molecule of A.

The reaction is activated by lactatedehydrogenase:

$$CH_3COCOOH + NADH + H^+ \longrightarrow CH_3CHOHCOOH$$

$$+ NAD^+$$

and may be represented as Fig. 7. According to ref. Fig. 7 the catalytic group for lactatedehydrogenase is the imidazole ring of histidine. The donor of the H^- is NADH. As in other cases of hydride ion transitions, the C_4-H bond in the pyridine ring, NADH, is not regarded as electrostatic, but in the elementary act of chemical change (in an active complex) the transition of hydride ion takes place.

Starting from the N^+ atom in the imidazole ring, the BRC for this process may be written in the form:

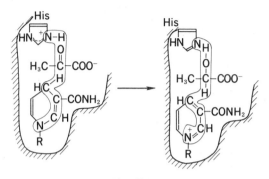

Fig. 7

(His) N⁺-H O=C H-C-C=C-N (Py)——→ N H-O-C-H C=C-C=N⁺

$\quad$ 1 0 2 0 1 1 2 1 $\qquad$ 0 1 1 1 0 2 1 2

$\Delta=\pm 1,\ 0.5$ $\qquad\qquad$ (1, 5; 1, 5; 1, 5)

This reaction is interesting in two respects. First of all, besides the system of atoms with alternating electronegativity, the system of conjugated bonds of the pyridine ring, NAD^+ is included in the BRC. The multiplicities of the bonds are 1 and 5, and for this part of the BRC the rule $\Delta=\pm0.5$ is valid. For the sake of the unification of this chemical mechanism, the simplest system of symbolization for the conjugated bonds was used. But the difference in the type of electron transport in chains of atoms with different electronegativity in the conjugated bond system makes it necessary to use different rules, namely $\Delta=\pm1$ and $\Delta=\pm0.5$, for different electron systems in the BRC.

Second, the transport of the proton from (and to) the enzyme and the hydride ion to (and from) the second participant of the reaction (NADH) makes it possible for the enzymes to act catalysis. The hydro-dehydrogenation process under the control of the proton transport to and from the catalytic groups of the active center. This means that the hydro-dehydrogenation here has the form of an acid-base reaction. The transport of H^+ and H^- in BRC lactatedehydrogenase it is possible to compare with two steps: the transition of H to pyruvate and the accompanying simultaneous redistribution of electron density in BRC.

(His)N$^+$-H O-C H-C-C=CN(Py)$\rightarrow$(His)N H-O$^+$=C H-C-C=C-NZ(Py)

(His)N H-O$^+$=C H-C-C=C-N(Py)$\rightarrow$(His)N H-O-C-H C=C-C=N$^+$(Py)

The charged carbon atom is trivalent and this is why the
intermediate products of the catalysis cannot be formed only
by accepting H$^-$ (or H$^+$), but have to change all their chemical
bonds in one elementary act.

Flavoproteins
    When the acid-base groups of enzyme molecule are
insufficient for the catalysis, the prosthetic groups of the
protein molecules are used. However, this is not a substitution for the acid-base catalytic groups, but acts in addition to them. In the active centers of all sufficiently
studied enzymes, at least one proton-donor or proton-acceptor group was found.
    The active centers of flavoproteins contain the acid or
basic groups as well as an isoalloxazine ring in the form of
FAD or FMN. If flavoproteins also contain metal ions, they
are regarded as metaloflavoproteins. All these enzymes
catalyze hydrogen transport by mechanisms including
hydride-ion transition conjugated with proton transition.
Acceptance of the hydride ion by flavines from donor DH$^-$
may use a different mechanisms, such as the transformation
of FAD to FADH$^-$. The latter is the stable form of FADH$_2$
in water solution at pH 7, since the equilibrium:

$$FADH_2 \rightleftharpoons FADH^- + H^+$$

corresponds to pK 6-7. The transport of H$^-$ has a BRC
action mechanisms shown in Fig. 8.

       N=C-C=N H$^-$-D $\longrightarrow$ N$^-$-C=C-N-H D
        2 1 2 0 1   1 2 1 1 0 $\Delta = \pm 1$

We have to note here that the real forms of the BRC are
slightly different, since they include also some part of the
hydride donor, molecule of D, changing by the hydride ion
transition. Metaloflavoproteins after accepting H$^-$ may
form one of the complexes shown in Fig. 9.
Here BH$^+$ is the proton donor-acceptor and B is the proton
acceptor in the protein molecule. The SH-group also belongs

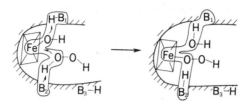

Fig. 8

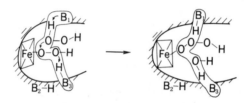

Fig. 9

to the protein molecule and is necessary for the formation of the metal-flavine complex and from carbon atoms of all substrates except NAD^+ ($NADP^+$). For example, the mechanism of the reaction realised by succinatedehydrogenase may be represented schematically as

$$
\begin{array}{ccc}
& COOH & \\
& | & \\
Fe{:}FAD{\cdot}HCH & \longrightarrow & Fe{\cdot}FADH^-{\cdot}CH \\
| \quad | & & | \\
B \quad HCH & & CH \\
| & & | \\
COOH & & COOH
\end{array}
$$

Here $FeFADH^-$ has the one form of the complexes (II, III or IV in Fig. 9) and the shortest form of the BRC is

$$(Fe{\cdot}FAD)\ H\text{-}C\text{-}C\text{-}H\ (B) \longrightarrow (Fe{\cdot}FADH^-)\ C{=}C\ H^+\text{-}(B)$$
$$\quad\quad 0\ \ 1\ \ 1\ \ 0 \quad\quad\quad\quad\quad\quad\quad 1\ \ 0\ 2\ 0\ \ 1$$

A regeneration of the active center of the succinatedehydrogenase is achieved by the successive accepting of two electrons and elimination of a proton:

$$Fe \cdot FADH^- \xrightarrow{\ e\ } Fe \cdot FADH \longrightarrow Fe \cdot FAD + H^+$$

In the reactions which concern H^- transport from electro-negative atoms N, O, S and from NADH, a hydride-ion intermediate acceptor is used. It is the disulfide group of the protein molecule. A mechanism of the reaction for lipoamidedehydrogenase may be schematically represented as follows:

$$NADH + S\text{-}S\ H^+\text{-}B \longrightarrow NAD^+\ H\text{-}S\ S\text{-}H \quad B$$

the second step being

$$B\ H\text{-}S\ S\text{-}H\ FAD \longrightarrow BH^+\ S\text{-}S\ H^-\ FAD$$

Here $FADH^-$ has the form of the complex I (Fig. 9) and B is the proton-acceptor. After these transformations the hydride ion is accepted by lipoamide with the help of second proton donor-acceptor in the active center of lipoamidedehydro-genase

$FADH^-$

$$\begin{array}{l} \overset{S\ \longrightarrow\ S}{CH_2 CH_2\text{-}CH_2\text{-}(CH_2)_4\text{-}CONH_2} \xrightarrow{\quad} FAD\ HS \qquad SH \\ \qquad +B \qquad\qquad\qquad\qquad\qquad\qquad CH_2\text{-}CH_2\text{-}CH_2\text{-} \\ (CH_2)_4\ \text{-}CONH_2 \end{array}$$

Here we note that all the steps in the flavine-accelerated reactions are linked to transitions in the different BRC, are controlled by proton transitions and possess all the pro-perties of the BRC mentioned above.

Hemoproteins

    The same general properties of the BRC may be found for hemoproteins. The simplest of these is catalase.[6,7] The active center of catalase contains heme and some acid-base groups. The valency of the iron (Fe^{3+}) is not changed during the decomposition of H_2O_2. The cooperative action of these catalytic groups makes possible the effective addition of the first molecule of H_2O_2 to the enzyme, the formation of an

instable intermediate with the second molecule of H_2O_2 and
the more rapid decomposition of H_2O_2 than would otherwise
occur.   The mechanism of these reactions was detailed by
us.[7]

   The first step of the reaction is shown in Fig. 10.   The
formation of hypothetical peroxide Fe-O-O-O-H corresponds
to the formation of Ogura complexes by the mechanism in
Fig. 11.   A decomposition of this unstable intermediate may
be described by a mechanism including the proton transitions
in the BRC (Fig. 12).

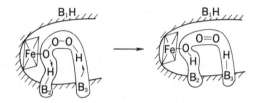

Fig. 10

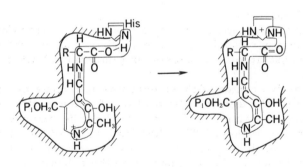

Fig. 11

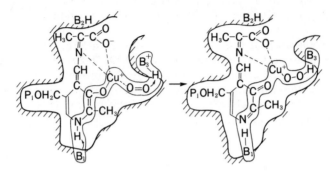

Fig. 12

From these examples it is possible to see that any next step in the enzymatic reaction will be linked to the transformation of oxidation-reduction processes, e.g.,

$$2H_2O_2 \longrightarrow 2H_2O + O_2$$

by the acid-base mechanism of catalysis in a linear BRC, inclucing the proton-donor and proton-acceptor steps.

Pyridoxal enzymes

    Numerous groups of enzymes of amino acid transformations contain pyridoxal as the prosthetic group, which in the enzyme-substrates complexes forms internal aldimines of the type.

    The active center of these enzymes also contains acid-base groups, the different dislocation of which on the body of the connected amino acids leads to the different ways of their transformations.[8] The chemical nature of the acid-base groups of the active centers of the pyridoxal enzymes is often not known. In this connection the schemes in which B and BH^+ are used regardless of the real charge of acid or base groups are given below. B may be $-COO^-$, $-S^-$, O^-(Tyr), $-NH_2$, or imidazole, while BH^+ may be $-COOH$, $-SH$, $-OH$ (Tyr), $-NH_3^+$, or the positive imidazolium ion. Here we are considering only reactions, but all the reactions data discussed in ref. 8 are concerned with the mechanisms, including the BRC, of proton transport.

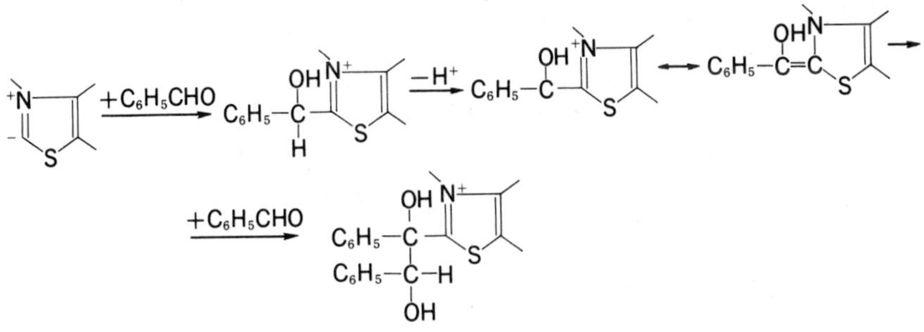

Fig. 13

The first step of the reaction, catalyzed by decarboxy-lases of amino acids, comprises only proton displacement. However, this creates changes in the multiplicities of the 10 bonds in the BRC, and is accompanied by CO elimination (Fig. 14).

Starting from the N atom of histidine in the active center, the total transition in the BRC may be represented in the form:

$$(\text{His})\text{N H-O-C-C-N=C-C=C-C=N}^{+}(\text{Py})--\text{N}^{+}\text{-H O=C C=N-C=C-C=C-N}$$
$$0\ 1\ 1\ 1\ 1\ 2\ 1\ 2\ 1\ 2 \qquad\qquad 1\ 0\ 20\ 2\ 1\ 2\ 1\ 2\ 1$$
$$\Delta = \pm 1$$

This example is important as the one-proton BRC. The chemical mechanism of the oxidation of amino acids by the pyridoxal-containing system proposed in ref. 9 is the same as that in the step of the scheme shown in Fig. 15 which involves the formation of hydrogen peroxide from O_2 .

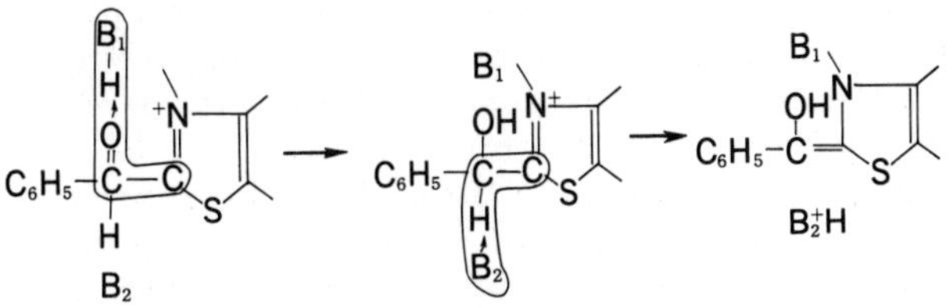

Fig. 14

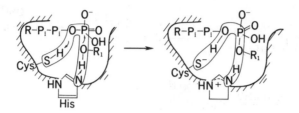

Fig. 15

Here it is possible to see that the role of the metal atoms in
enzymatic catalysis does not differ greatly from the role of
the other atoms in the BRC, and is concerned in this case
only with the possibility of forming a chemical bond with
oxygen. The valency of Cu in the BRC does not change
throughout all the stages of the reaction. The BRC transitions are linked to two proton transitions which are performed with the assistance of acid and base groups of the active
center:

$$B_1 \; H\text{-}N\text{-}C\text{=}C\text{-}O\text{-}Cu^+ \; O\text{=}O \; H^+\text{-}B_3 \longrightarrow B_1\text{-}H \; N\text{=}C\text{-}C\text{=}O \; Cu^+\text{-}O\text{-}O\text{-}H \; B_3$$
$$\;\; 0\,1 \quad\;\; 1\; 2\; 1\; 1 \quad\; 0 \;\; 2\; 0\; 1 \qquad\qquad 1\; 0\; 2\; 1\; 2\; 0 \quad\; 1 \quad 1\; 1\; 0$$

For these reactions, as in other cases, it is possible to
recognize stepwise changes in the BRC due to the subsequent
transition of the protons and resulting a redistribution of the
remaining covalent bonds.

Enzymes Containing DTP
     The material given above confirms the idea that the
presence in the enzyme globula of the prosthetic group does
not essentially influence the reaction mechanism based on the
transitions in the BRC, linked to the proton displacements.
As a final example we turn to the properties of the DTP-
containing enzymes. The general mechanisms of these were
proposed by Breslow. The mechanism of the benzoin condensation discussed by B. Pullman[10] may be represented in
Fig. 16. Another mechanism of the same step involves
two acid-base groups of the protein molecule (Fig. 17).
     For each step of this reaction there is a BRC the transi-
tion of which is governed by the proton movements to and

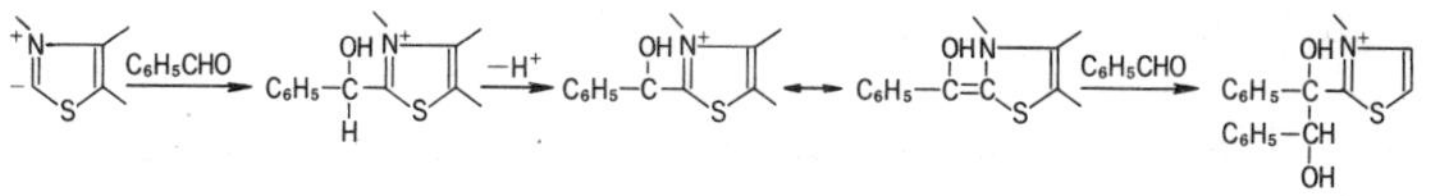

Fig. 16

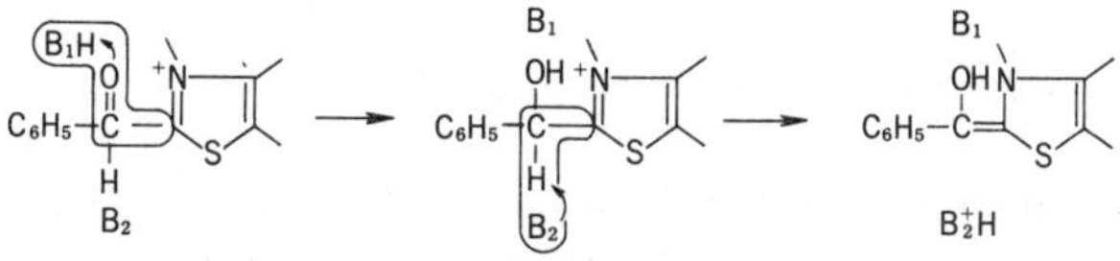

Fig. 17

from the catalytic groups. In connection with this mechanism we proposed that the driving force of the enzymatic reaction is the gain energy of resonance in the transition state. We do not agree as the disappearence of the system of the conjugated bonds. Secondly, a lot of enzymatic reactions do not contain systems of conjugated bond systems in their BRC at all, but all of them have common properties. The BRC is always governed by one or two proton exchanges between electronegative atoms.

Some properties of the BRC of Enzymatic Catalysis

The chemical changes always involve some systems with changes in bond multiplicities. The simplest chemical reaction gives the first example:

$$D^+ + H\text{-}H \longrightarrow D\text{-}H + H \qquad E_{act} \approx 8 \text{ kcal/mole} \qquad (5)$$
$$ 0 \;\; 1 \phantom{\longrightarrow D\text{-}H +} 1 \;\; 0$$

or

$$D^+ + H\text{-}H \longrightarrow D\text{-}H + H^+ \qquad E_{act} \approx 0 \qquad (6)$$
$$ 0 \;\; 1 \phantom{\longrightarrow D\text{-}H +} 1 \;\; 0$$

Of course not all the elementary acts include a BRC. The recombination process is an example:

$$D + H + H \longrightarrow D\text{-}H + H$$
$$\quad 0 \quad 0 \qquad\qquad 1 \quad 0$$

or excitation by the stroke of the second order. However, the chemical changes usually involve BRCs. For the enzymatic catalysis very important is the act of A general basic catalystic reaction in the presence of the proton acceptor B is very important for the enzymobysis (Fig. 18).

It is easy to show the four-ring BRC:

$$B \text{ H-O C=O} \longrightarrow B^+\text{-H O-C-O}^-$$

The reason for the chemical change in the form of BRC is explained by quantum mechanical treatment on the quoted example of this type of reaction.[5] The breaking of some bonds proceeds unhinderd if a new chemical bond is formed near them. This process has the form of a single elementary act, with the redistribution of electron density corresponding to the alternating increasing and decreasing of the multiplicities of the bonds. The general theory refers only to the processes of the type[5,6] but the same result is valid qualitatively for many chemical processes.

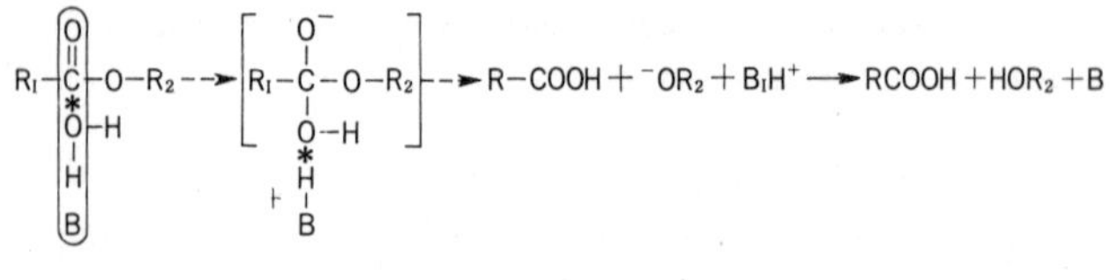

Fig. 18

We must emphasize that the enzymatic and nonenzymatic catalysis do not differ in their BRC formation. Their difference is only in their BRC properties. Regarding the mechanism of the enzymatic reactions given above, we may conclude that for the majority of enzymatic reactions the BRCs have the form of the sum of two BRCs of the simple reactions of the proton or hydride ion transport. Very seldom, as in the case of pyridoxal racemases, the enzymatic BRC corresponds only to a single proton transition. The typical enzymatic BRC is formed by two catalytic groups which attack the changeable bond of the substrate:

Catalytic	Changeable bond	Catalytic
group I	or bonds of substrate	group II

This system also possesses the properties necessary for the push-pull mechanism of the enzymatic reaction, because the catalytic groups I and II in all cases have different acceptor-donor properties.

However, this conclusion has to be supported by the supplementary data, because the stepwise transition in the BRC cannot yet be excluded. The reactions (2)-(4) are given only to show this possibility for α-chymotrypsin.

We must note firstly that the construction of the combined BRC in enzyme globula is constructed from catalytic groups and parts of the prosthetic groups, while necessary bonds in the substrates give rise to the very important differences in the performance of various enzymatic reactions. The combination of the bonds in the BRC, depending only on the interaction of the participants of the catalytic processes with the enzyme globula, makes possible the construction of the unique mechanism of the reaction, which can never be simulated in solution. In this case the catalytic groups have only one way for attack, there being no possibility of the substrates and intermediate products taking part in the rapid competition processes. The sequence of the chemical changes is governed by the type of enzymatic BRC. All this leads to the qualitatively new possibilities in enzymatic catalysis. The most interesting examples here are connected with the acid-base mechanism for many of the enzymatic reactions. In homogeneous systems the same reaction may have different mechanisms not connected with the acid-base catalysis.

In hematin solution a decomposition of H_2O_2 is not connected with the transitions of protons. The active center of catalase contains heme and three acid-base groups (see above). All the processes in catalase are ruled by the displacements of two protons for each of three BRCs during the decomposition of $2H_2O_2$ to $2H_2O$ and O_2. The impossibility of creating the BRCs of catalase in solution leads to the change of the chemical mechanism of the reaction and to the decrease by two to three orders in the catalytic efficiency. In many other homogeneous catalytic reactions of oxidation-reduction type, which are not essentially acid-base, in the enzymatic BRC. Naturally it is not possible in all the oxidation-reduction processes possible to effect the enzymatic BRC by only

two protons transitions. In such cases the proton and hydride
ion displacements in the BRC have been studied. A review
of the available material shows that it is sufficient for the
performances of most of the oxidation-reduction processes
which are interesting to biochemists by virtue of their acid-
base catalytic mechanisms.

The combination of the two specially selected catalytic
acts and the possibility to inclusion in the BRC of only the
bonds of the substrates is the usual way of creating of new
mechanisms of artificial enzymatic catalysis.

Kinases transfer the phosphoryl groups from ATP to the
different substrates rather than more straightforward processes, such as the hydrolysis of ATP only because the substrates are included in the corresponding BRC, but not in the
molecule of water. An example of this is the reaction
catalysed by hexokinase (Fig. 19). Here R_1OH is hexose.
The six-membered BRC here has four constituents, two parts
of the catalytic groups and the two reactive bonds of the two
substrates. Because of its low energy of activation, this
four-molecular reaction is highly improbable in solution.

The construction of the BRCs of kinases is connected
with the many groups of the adsorption center of the enzyme
globula and by the globula's tertiary structure. This is the
way they influence the chemical mechanisms of the enzymatic
catalysis.

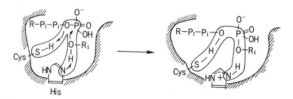

Fig. 19

We must emphasize that the formation of the BRC in
homogeneous systems is dependent only on the geometrical
and energy properties of the participants in the BRC - the
catalytic groups reactants. In the enzymatic globula it
depends to a great extent on the interaction of those same
participants with the protein structure. This corresponds
to other principles of the organization of catalysis and opens
up new possibilities for our understanding of the selection

and mutual orientation of the bonds of enzymatic reactions'
BRCs.

In contrast to homogeneous systems, in enzymatic
globulas it is possible for different polar surroundings to be
established in different parts of the BRC. This leads to an
increase in the necessary acceptor and donor properties of
the catalytic groups. In such cases the same BRC according
to chemical composition possess different properties in
solution and in the globula.

A specific orientation of the substrate in enzyme-
substrate complexes is well known in enzymology. In ref. 11
the energetic aspect of the bond orientation in the BRC is
discussed. The change of the bonds' orientation leads to a
change in the energies of the binding of the catalytic groups
with the substrate.[11] However, another aspect of the same
question concerns the change in the mechanism of catalysis
due to the fixation of catalytic groups and substrates in the
protein globula. For example, in many cases of enzymatic
catalysis a proton acceptor has only indirect contact with the
substrate molecule — through the water molecule. For the
imidazole in active centers, this leads to its acting on the
mechanism of general base catalysis, instead the nucleophilic
catalysis, as usual for aqueous solutions (Fig. 20).

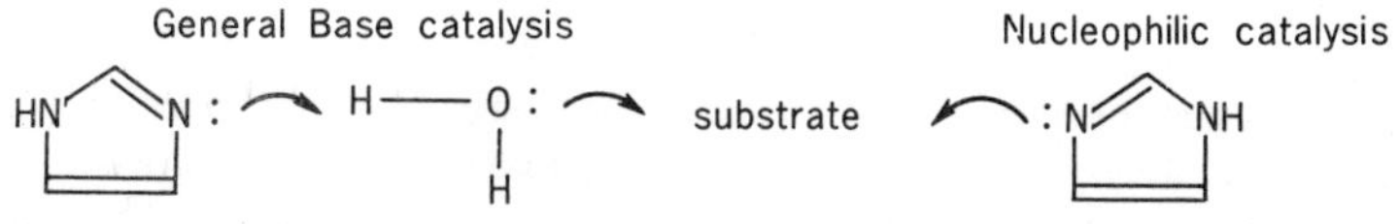

Fig. 20

Now we must consider the mechanism of the transitions
in the BRCs. The fact that proton displacements in BRCs
may often be considered separately does not mean that these
transitions are proceeding stepwise rather than as a push-
pull mechanism. It reflects rather the evolution of the
catalytic process organised in enzymatic globula by selection
of the necessary homogeneous reactions. The combined
BRCs of the enzymes are constructed from the simpler parts
of the BRC for single proton transitions. Although it is not
possible to make any general conclusions, it does often

occur that single-proton transitions proceed more slowly
than the integrated enzymatic reaction. If the energy of
activation of any single step is more than that for the BRC
of the complete reaction, this would prove the push-pull
mechanism of the transition for enzymatic BRC. In this
case enzymatic catalysis has a low energy of activation of a
polymolecular process and high pre-exponent of a bimolecular
reaction. The simultaneous action of acceptor and donor on
the same substrate in the enzymatic BRC makes both proton
and hydride ion displacements easier. The reason
for this is the same as for the facilitation of the formation
of a new bond when a neighboring bond is broken and a cooperative exchange of electron densities occurs. Therefore
we expect that transitions in BRC's have the push-pull mechanism. However, precise evidence for this statement has
been obtained for only a few cases.

We have now to discuss the properties of the enzymatic
BRCs as physical systems. Most interesting is the physical
sense of the rule ± 1. A reorganization of the covalent bond
system is always in the form of electron density redistribution.
The initiation of this process by means of heterolytic hydrogen
transport to or from the catalytic group in an electroneutral
system necessitates electron transport in the BRC. The
general condition of the conductivity of the BRC has the form:

$$\text{any } \Delta \neq 0$$

where Δ is the change in the multiplicities of the chemical
bonds in the BRC. The electron transport in the BRC is
possible only in an unstable chemical bond system, even one
stable covalent bond in the BRC blocks the electron transport.
In the sense of conductivity, the BRC is an analog of the
organic semiconductor. The condition for facilitation of the
bond redistribution has the form:

$$\Delta = \pm a$$

Finally, the use in enzyme groups of univalent transport
particles (H^+ or H^-) gives the condition:

$$\Delta = \pm 1$$

which is valid for the heterolytic mechanism of bond redistribution. However, this does not mean a one-electron

process. When H transition occurs in the BRC, two electrons are transferred for the formation of one covalent bond. In other parts of the BRC one-electron transport takes place. In this case the atoms involved in the BRC change in their electronegativities and present an alternative electric field in the system. When a nonpolar system of conjugated bonds is present in the BRC, a new rule for the transitions in the BRCs occurs:

$$\Delta = \pm 0.5$$

This corresponds to other physical mechanisms of electron transport in these systems.

The phenomenology of the enzymatic BRC is considered in ref. 1. Here we need only consider link of the transitions in the BRCs with the shift in the total electrical charge. From this point of view we have four types of BRCs, which may be represented as Fig. 21.

In the BRCs may occur the transport of positive or negative charge, the division and combination of the positive and negative charges, and their redistribution in the cycle. Corresponding examples are given in Table I.

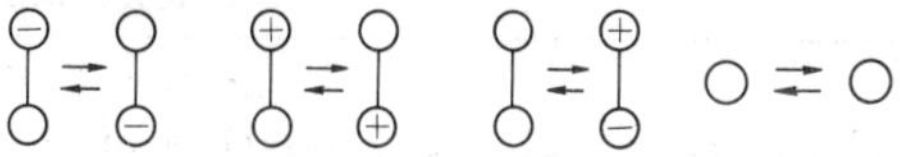

Fig. 21

The type of the charge redistribution in the BRC is connected with the composition of the active centers of enzymes, but it does not define the mechanism of catalysis. Among the catalytic groups of the protein molecules the proton donors are either positively charged (imidazolium ion, $-NH_3^+$) or neutral ($-COOH$, $-SH$, phenol $-OH$). The acceptors are either negative ($-COO^-$, $-S^-$, -0^-) or neutral (imidazole, $-NH_2$). Any BRC contains either acceptor and donor or only one catalytic group. The first case gives the following three combinations for the charge transport:

His - His$^+$

COOH - COO$^-$

His$^+$ - COO$^-$

and the redistribution of the charge in the BRC cycle. The transitions in one-proton BRCs are always connected with the division of the charges (and reverse process). However, the type of the charge transport in BRC does not define the mechanism of catalysis, but rather the reverse is true. In the BRC of α-chymotrypsin described by Blow[4] the transport

Table I. Redistribution of the electrical charge in the BRCs.

Type	Form of the heterolitic hydrogen transport	Number of the broken and reformed bonds	Enzymes and number of the corresponding members in BRC
Shift in the negative charge	$2H^+$	$3 \rightleftharpoons 3$	α-Chymotrypsin (8) Carboxypeptidase (6) Lysozyme II (6)
Shift in the positive charge	$2H^+$	$3 \rightleftharpoons 3$ $2 \rightleftharpoons 2$	Ribonuclease (6) Transaminases (6) Alaninoxydase (10)
Shift in the positive charge	H^+ and H^-	$2 \rightleftharpoons 2$	Lactatedehydrogenase (8) NAD^+ and $NADP^+$ Dehydrogenases
Shift in the positive charge	H^+	$1 \rightleftharpoons 1$	Pyridoxal decarboxylases of amino acids, II stage (8) Treoninaldolase Serinedehydrotase
		$1 \rightleftharpoons 2$	Pyridoxal decarboxylases of amino acids, I stage (10) Treoninaldolase
Divisions of the charges	$2H^+$	$3 \rightleftharpoons 3$	Hexokinase (6) Acetylcholinesterase (6)
Divisions of the charges	H^+	$1 \rightleftharpoons 1$	Racemases of amino acids (2) Esterase (6)
Divisions of the charges	H^+	$2 \rightleftharpoons 2$	Galactosidase

of negative charge occurs. For the same enzyme may take place by the mechanism of Bender the redistribution of charge by cycle. In the BRC of ribonuclease it is the positive charge which is shifted. In the BRC of acetylcholinesterase we have division of charges. But in all these cases the catalytic reaction mechanisms are similar. The question is the BRCs with two-proton transitions ruled by the proton acceptors and donors as catalytic groups of enzymes. The chemical nature of the acceptors and donors is different, and this is the cause of the observable difference in charge transport.

The mechanisms of the enzymatic reaction well-known at the present time correspond only to three types of BRCs. These are systems with two-proton displacements, with displacement of H^+ and H^- and with the transition of one proton. By their general construction, enzymatic reactions the BRCs are also similar; in fact there are only two types. Two catalytic groups (acceptor and donor) are attacking from opposite sides the reactive bond (bonds) in a substrate, or near the reacting bond is placed one catalytic group and a system of conjugated bonds able to accept charge (e.g. pyridine ring).

The reason for the transitions in the BRCs is connected with the transfer of proton (protons) to or from the catalytic groups. The changes in the state of the BRC may not be regarded simply as the results of the electron transport in the BRC. The latter is the consequence rather than the reason for the transitions in the BRCs. However, detailed chemical differentiation is not possible here because we have to deal with an unobservable phenomenon which, according to quantum mechanics, does not make real physical sense.

The main conclusion from this analysis of the mechanisms of enzymatic reaction may be summarized as follows. The difference in the mechanisms of homogeneous and enzymatic reactions is connected with the possibility of picking up the sequence of bonds in the BRC in the protein globula. This may make possible the creation of new reaction mechanisms (especially for oxidation-reduction processes). In the enzymatic globula any act of catalysis may be possible in optimal conditions, which depend on the choice of ligands and the removal of concurrent processes. The combination in enzymatic BRCs of pairs of relatively slow catalytic processes with a push-pull mechanism may make both of the pair more effective. All these properties together result in the qualitative differences between enzymatic and homogeneous catalytic reactions. The second task of this paper was to illustrate the potentials of the BRC concept for future analyses of rather complicated mechanisms of enzymatic catalysis. Besides its other advantages, this approach helps our understanding of the similar catalytic mechanisms of many apparently different types of reactions.

This study has not taken into account the selectivity of enzymatic reactions. The chemical approach and chemical mechanisms of the enzymatic reaction are insufficient for the analysis of this problem.

References

1) O.M. Poltorak, Vestn. Mosk. Un-ta, ser. khimii, no. 12, 642 (1971).
2) O.M. Poltorak and E.S. Chukhrai, "The Physico-chemical Aspects of Enzymology," M. Visshai shkola, ed., 1971.
3) D. Phyllips, "The molecules and cells," M. Mir, ed., 1968.
4) D.M. Blow, J.J. Birktoff and B.S. Hartley, Nature, 221, 337 (1969).
5) W.N. Lipscomb, Accounts of Chem. Res., 3, 81 (1970).
6) P. Jons and A. Suggett, Biochem, J., 110, 621 (1968).
7) O.M. Poltorak, Vestn. Mosk. Unta, ser. khimii, no. 12, 3 (1971).
8) O.M. Poltorak and E.S. Chukhrai, Vestn. Mosk. un-ta, ser. khimmi, no. 12.
9) "Chemistry and biology of pyridoxal catalysis M.," Nauka, ed., 1968, p. 321.
10) B. Pullman, "Chemical and Biological Aspects of Pyridoxal Catalysis," Pergamon Press, Oxford, London-New York-Paris, 1963, p. 133.
11) O.M. Poltorak, Vestn. Mosk. un-ta, ser, khimii, no. 6, 15 (1968).

Discussion

A. P. Purmal

I would like to direct a question to Professor Saito: Do two electron transfers really take place?

Y. Saito

The step of hydroxymercuration is a simple addition reaction not followed by an electron transfer. It is rather like hydration. Indeed, if we add hydrochloric acid the reverse reaction-deoxymercuration-proceeds, yielding olefin. Hydroxymercurated olefin contains Hg (II), not mercurous but mercuric.

Since no kinetic dependence on free Hg^{2+} was found, I believe the reaction proceeds in a monomolecular way that indicates two electron transfers. The homolytic radical splitting does not take place

$$CH_3\overset{\frown}{C}H\!\!-\!\!CHCH \longrightarrow CH_3\overset{\|}{\underset{O}{C}} CH_2CH_3 + Hg\ (O) + H^+$$
$$\underset{OH}{|} \quad \overset{\wedge}{Hg^+}$$

But after the reaction took place in the solution, with the help of Laser Raman Spectroscopy I found Hg^{2+}. It results from a rapid equilibrium reaction, $Hg^{2+} + Hg\ (O) \rightleftharpoons Hg_2^{2+}$, with an equilibrium constant preferred to Hg^{2+}.

K. I. Matveev

The results presented by Professor Saito help very much to elucidate the mechanism of olefin oxidation reaction not only in the presence of mercury salts (II) but in the presence of palladium compounds as well. From this work it follows that mercury (II) and palladium (II) salts catalyze the reaction following the same mechanisms. Professor Saito assumes that in the case of palladium salts (II) the rate-determining step is the σ-complex formation. This assumption is based on the Henry hypothesis. We have analyzed the data on which this hypothesis is based. The experiment is in better agreement with the assumption that the oxidation of ethylene as well as of other olefins in the presence of Pd (II) complexes is limited not by the

σ-complex formation but by its redox decomposition, similar to that in the presence of mercury salts (II).

Extremely important is the conclusion by Professor Saito that the catalytic property changes are connected with the changes of both acid-base and redox properties of metal ions. It seems to me that now it is necessary to try to establish a quantitative correlation between these alterations.

K. I. Matveev

I would like to direct a question to Professor Saito: Do you have your own experimental data which gives evidence that the rate-determining step for Tl^{3+} oxidation is the formation of a Π-complex?

Y. Saito

The conclusion on Tl^{3+} oxidation was given by Henry. Now our research group is carrying out the investigation, but at present I cannot tell you about it. To explain it, I would like to inform you about the relationship between the kinds of metal ions and the position of a rate-determining step as regards a chemical bond. Generally, organomercuric compounds are stable, but organothallic compounds are not as stable. Indeed, the oxythallation product of styrene is obtained as a σ-complex, but it is rather easily decomposed. That is why I believe that the Tl^{3+}-C bond is relatively weak. It seems reasonable to me that the rate-determining step is the Π-complex formation for Tl^{3+} oxidation.

Y. Saito

The reason I investigate homogeneous catalytic systems by the NMR method is that this method gives direct information on the chemical bond between substrates and catalysts. Indeed, the values of ^2IHgH of hydroxymercurated olefins show the extent of σ-bond between C and Hg (II). When we change an anion from ClO_4^- to $OCOCF_3^-$, I-values and the rates of redox decomposition decrease. I should say that time-sequential NMR spectroscopy is useful in studying the kinetic isotopic effect. Actually, in the present investigation I could get a normal (not inverse) secondary deuterium isotopic effect and a very small primary isotopic effect. That is why I represent the activated complex as

$$\left[\begin{array}{ccc} H_3C & H & CH_3 \\ & C\!\!-\!\!-\!\!-\!\!-\!\!C & \\ Ho & Hg & H \end{array}\right]^{+}$$

For quantum-chemical calculations using the EMO approach it is necessary to know the nature of the chemical bond of this activated complex as well as that of reactant mercurials.

I believe that from the chemical bond viewpoint there is no difference between the homogeneous and heterogeneous catalysis. Allyl intermediates on the surface of selective oxidation catalysts or intermediates in the reaction of olefin disproportionation are very interesting for me from this viewpoint.

V. V. Malakhov

In the report by Professor Sokolski it was shown that electrochemical methods permitted very important information on the metallic catalyst surface state to be easily obtained. At the Institute of Catalysis we studied the dependence of the electroconductivity of powders of some metals (cuprum, silver, nickel, ferrum) on the value of the electric field strength. In the case when oxygen, sulfur dioxide, ethylene, acetylene, divinyl, and cyclohexane were adsorbed on the surface of these metals the powders did not conduct electric current at low voltage values. However, the voltage increase resulted in an electric gap in the system. The gap is the result of the spatial structure formation by a dispersed metal in an electric field, after which this structure becomes a current conductive bridge. As it proved to be, other conditions being equal the value of a gap voltage is determined by the nature of a substance adsorbed on the metal surface. For example, in the case of nickel the gap voltage value was for SO_2: 55 V, O_2: 95 V, acetylene: 100 V, ethylene: 120 V, and divinyl: 130 V under the conditions of a measuring cell.

Having reported on the effect found, we believe that the gap voltage value as the energetic parameter of the metal powder surface which, being easily determined, can be useful in studying catalysis and adsorption phenomena.

T. Seiyama

I am greatly interested in the work presented by
Professor Sokolski. The use of an electrochemical technique to investigate catalysts was shown in hydrogenation
reactions. To my mind it is a promising technique.
Unfortunately, the report is made in a form too condensed
to allow detailed understanding, especially as concerns the
experimental procedure. Now I am working on oxidation
catalysts and believe that the electrochemical technique of
Professor Sokolski can be applied to some cases of catalytic
oxidation. In my study I found that Pd-black suspended in
water acts as a catalyst of propylene oxidation to acrylic
acid, contrary to the Wacker process in which Pd ions
oxidize propylene to acetone (J. Catalysis, in press).
About the oxidation by Pd-black little is known due to the
difficulties in obtaining data on adsorption of propylene,
oxygen, etc. If it is possible to apply this technique much
information on catalytic oxidation will be obtained. I would
like some suggestions on using the electrochemical technique to study the oxidation catalyst.

K. Tamaru

In connection with a very interesting speech by Professor
Shylov I would like to report briefly some results we have
obtained on ammonia synthesis.

Neither transition metal phthalocyanines nor sodium (nor
potassium) metal chemisorbs nitrogen and hydrogen.
However, when they are brought into contact to form the
EDA complexes they begin to chemisorb hydrogen and
nitrogen and to catalyze ammonia and Fischer-Tropsch
synthesis, hydrogenation, and H_2-D_2 exchange reaction.
The catalytic activity markedly depends on the negativity of
the phthalocyanine anion in the complex. A more highly
negative phthalocyanine anion exhibits higher activity.

When the EDA complex of Fe-phthalocyanine with sodium
is deposited on activated charcoal with a large surface area,
it catalyzes the ammonia synthesis to an appreciable extent
even at room temperature. Another type of the EDA complex for ammonia synthesis is that of graphite with potassium. It chemisorbs hydrogen and catalyzes hydrogenation
and H_2-D_2 exchange reaction.

In particular, if the transition metal chlorides such as
$FeCl_3$ are added to the graphite-potassium system, the

catalytic activity significantly increases. (The preparation
of this catalyst is the following: Graphite + $FeCl_3 \longrightarrow$ inter-
layer compound $\xrightarrow{+K(excess)}$ catalyst $\longrightarrow$ evacuation at 100-
150°C to remove excess potassium metal.)

A. B. Shylov
  The problems discussed in the report of Professor O. M.
Poltorak and E. S. Chukhrai are very important. But it is
believed that the conclusion made in the report that the
mechanism of a great number of fermentative reactions is
absolutely clear is too optimistic. Really it may concern
the simplest cases for which in addition nonfermentative
analogues are known. By the way, for these analogues it is
quite possible to apply the rule of the bond order change by
±1. In particular, for the electrophilic substitution in
benzene ring, for example, in the case of aniline:

1 2 1 2 0   2 1 2 1 1

or in Cannizzaro reaction:

1 0 2 0 1 1   0 1 1 1 0 2

  It is possible that while drawing a conclusion on the
mechanism of a fermentative reaction the authors always
take into account the mechanism of a nonfermentative analogue. If such an analogue is unknown (as it was until
recently for the fermentative nitrogen fixation or hydrocarbon oxidation) the building up of a true mechanism of a
fermentative reaction is practically impossible.

O. M. Poltorak

First of all I would like to touch on the question of
"simplicity" of the fermentative catalysis mechanisms.
Can it be said without exaggeration as it was in the report
that the fermentative catalysis mechanisms are uniform and
relatively simple? I think it cannot, provided that some
necessary reservations are given. It can be said that some
general laws of nature are simple, but not that they are
easy and simple to investigate. The study of the fermentative catalysis mechanisms would require vast labor and
time consumption, detailed and long-term investigations.
In addition, the possibility of pointing to sufficiently true
mechanisms of four fermentative catalysis reactions has
appeared only as a result of X-ray structural analysis of
proteins that to a certain extent is the summit of experimental art. Such data on mutual arrangement of reagents and
all groups in contact with them can be found in no other case
of homogeneous or heterogeneous catalysis. These four
ferments relate to the most completely studied catalytic
systems. For other ferments the catalysis mechanisms are
not yet known with such assurance and were obtained by way
of a detailed, long-term, and difficult investigation. The
result proved to be quite unexpected: the mechanisms
suggested by various authors and at varying times proved
to be reasonably nearer to each other than the authors
themselves supposed and than could be anticipated from
general considerations by analogy with the mechanisms of
homogeneous and heterogeneous catalysis. It is this idea
that is in the basis of the approach for the analysis of
fermentative catalysis mechanisms given in the report.
What is more, all the processes considered can be placed
into three subclasses of one class. It is said about the
functioning of linear succession of bonds with the transfer of
one proton, two protons or a proton and a hydride ion in the
catalysis act. It significantly decereases the number of
hypothetical mechanisms discussed in the analysis of one or
another fermentative reaction. Certainly, the possibility
cannot be eliminated that among many insufficiently studied
reactions there are processes following other mechanisms.
Still I believe that the number of such processes, if any, is
relatively small.

The second question which should be dwelled on concerns
the physical meaning of the "± 1" rule while describing
functions of BOC (bond overdistribution circuit). In many

simple cases such as removals in the conjugate bond system
the "±1" rule inevitably appears and seems almost trivial.
However, we consider more complex problems, and the
discussion of catalysis mechanisms is not restricted only to
the examination of valence schemes, universally determined
by general chemical laws. Here I should first of all point
to the bond overdistributions in a group of chemically
unbound molecules. As an example I could refer to the
hexogenase ferment substrate complex shown in the report.
In this case a six-member circuit of overdistributed bonds
includes the fragments of four molecules — two catalytic
groups and two substrate molecules. Without disturbing the
laws of nature one can imagine several methods of phosphyl
group transfer without building up a single system of BOC
relations, and in some earlier works such hypotheses were
actually advanced. It should be also noted that the act of
fermentative catalysis could be described not only by a
single circuit, the behaviour of which is regulated by the
"±1" rule, but also, for example, by two scraps of such a
circuit divided by a group of chemically unchangeable bonds
or even one such bond. It would immediately exclude the
"±1" rule from the range of laws of nature. But in either
known case this did not take place. The latter suggests an
idea that the building up of BOC and the "±1" rule describes
a general and far from self-evident feature of the fermentative reaction mechanisms. Certainly, there are many cases
when the "±1" rule refers to simple and even trivial fertures
of the construction of molecules studied or model systems.
Still it seems to me that this should not compromise the
importance of the rule considered in the fermentative catalysis.

And finally I would like to dwell on different mechanisms
of homogeneous and fermentative catalysis. On the basis of
the fact that with the help of the same principles it can be
possible to analyze the mechanisms of homogeneous and
fermentative catalysis, it does not follow that for the same
reaction general mechanisms of homogeneous and fermentative reactions will coincide or will even be common. In
fact, such coincidence does not generally exist, but the
reasons for it are not quite clear. In a ferment-substrate
complex the catalytic groups of protein and substrate molecules are kept in conformations favorable for building up
BOC by numerous groups of the adsorption center or, as it
can be said, are supported by the whole tertiary structure

of a protein globule. It greatly affects the catalysis action
itself, as in the ferment not only favorable for this orientation of reacting bonds is realized, but also the necessary
nearest order around each of the reacting groups is created,
the competition of solvent molecules is eliminated, and
thereby necessary chemical properties of reagents are
formed. As it is known in the case of lyzozyme, two carboxyl groups of the scrap active center $A_{SP} = 52$ and $Glu = 35$
have reasonably different chemical properties: one is an
active acceptor and the other an active proton donor. For a
homogeneous group this is impossible. Such phenomena
greatly hinder the modelling of fermentative reactions and
at the same time they permit us to understand the reasons
for varied chemical properties manifested in ferments by a
relatively small number of acid-base groups of protein
molecule active centers.

All this permits us to note in conclusion that the principles determining the course of elementary acts of homogeneous catalytic reactions and fermentative processes are
most probably general, and all those large differences which
are observed in activities, selectivities, and mechanisms of
the catalytic action of homogeneous catalysts and ferments
are determined only by the realization of catalytic "cells",
which differ first of all in their structural properties and
chemical composition, i.e., in the selection of catalytic
groups, or simply in their nearest surroundings which are
formed in homogeneous solutions and ferment globules in
absolutely different ways.

A. P. Purmal
I would like to make a few concluding remarks after the
comment: I would like to note that in the work on the
elucidation of the role of conjugation the acid-base and
redox stages in a catalytic act, the scientists of the
Novosibirsk Scientific Center — V. M. Berdnikov, a
member of the Institute of Kinetics and Combustion, and
L. N. Arzamaskova, a member of the Institute of Catalysis,
take part. I wish to thank the Organizing Committee of the
1 st-Soviet-Japanese Seminar for the honor given to me of com-
pleting fruitful work with my report.